The Consistency of Arithmetic

The Consistency of Arithmetic

AND OTHER ESSAYS

Storrs McCall

OXFORD
UNIVERSITY PRESS

OXFORD
UNIVERSITY PRESS

Oxford University Press is a department of the University of Oxford. It furthers the University's objective of excellence in research, scholarship, and education by publishing worldwide.

Oxford New York
Auckland Cape Town Dar es Salaam Hong Kong Karachi
Kuala Lumpur Madrid Melbourne Mexico City Nairobi
New Delhi Shanghai Taipei Toronto

With offices in
Argentina Austria Brazil Chile Czech Republic France Greece
Guatemala Hungary Italy Japan Poland Portugal Singapore
South Korea Switzerland Thailand Turkey Ukraine Vietnam

Published in the United States of America by
Oxford University Press
198 Madison Avenue, New York, NY 10016

Library of Congress Cataloging-in-Publication Data
McCall, Storrs.
The consistency of arithmetic : and other essays / Storrs McCall.
pages cm
Includes index.
ISBN 978–0–19–931654–0 (alk. paper)
1. Philosophy, Canadian—20th century. I. Title.
B995.M331 2013
191—dc23
2013008563

1 3 5 7 9 8 6 4 2
Printed in the United States of America
on acid-free paper

{ CONTENTS }

{ PROVENANCE OF THE ESSAYS }

Can a Turing machine know that the Gödel sentence is true? *The Journal of Philosophy* 96.10 (1999), pp. 525–532.

On "seeing" the truth of the Gödel sentence. *Facta Philosophica* 3.2 (2001), pp. 25–29.

How to make a decision. In *Action, Rationality and Decision*, ed. Vanderveken and Fisette. London: College Publications (2008), pp. 325–338.

Indeterminist free will, by Storrs McCall and E.J. Lowe. *Philosophy and Phenomenological Research* 70.3 (2005), pp. 681–89.

The determinists have run out of luck, for a good reason, by Storrs McCall and E.J. Lowe. *Philosophy and Phenomenological Research* 77.3 (2008), pp. 745–748.

The open future and its exploitation by rational agents. In *Around the tree: Semantic and metaphysical issues concerning branching and the open future*, ed. Correia and Iacona. Dordrecht: Springer (2012), pp. 95–104.

Does the brain lead the mind? *Philosophy and Phenomenological Research* 86.2 (2013), pp. 262–265.

3D/4D equivalence, the twins paradox, and absolute time, by Storrs McCall and E.J. Lowe. *Analysis* 63.2 (2003), pp. 114–123.

Philosophical consequences of the twins paradox. *The Ontology of Spacetime*, ed. D. Dieks. Amsterdam: Elsevier (2006), pp. 191–204.

The 3D/4D controversy: A storm in a teacup, by Storrs McCall and E.J. Lowe. *Nous* 40.3 (2006), pp. 570–78.

Laws of nature and branching spacetime. Paper presented at the annual meeting of the Society for Exact Philosophy, Winnipeg, Manitoba, May 2011.

Objective time flow. *Philosophy of Science* 43.3 (1976), pp. 337–362.

Time flow. In *The Importance of Time*, ed. L.N. Oaklander. Dordrecht: Kluwer Academic Publishers (2001), pp. 143–151.

QM and STR: The combining of quantum mechanics and relativity theory. *Philosophy of Science (Proceedings)* 67 (2000), pp. S535–S548.

The supervenience of truth: freewill and omniscience. *Analysis* 71.3 (2011), pp. 501–506.

An insoluble problem. *Analysis* 70.4 (2010), pp. 647–48.

The Consistency of Arithmetic

Introduction: The Consistency of Arithmetic and Other Essays

1. The Different Themes of the Book

A dozen different philosophical themes are reflected in the essays published here. The first, and most controversial, is discussed below and in essay 1, The Consistency of Arithmetic. Essays 2 and 3 deal with the relationship between provability and truth. The third group, essays 4, 5 and 6, deal with practical deliberation, Aristotle's *bouleusis*. This concerns our ability, as agents, to freely choose which one out of a set of different possible alternative actions to perform. The same theme is continued in the fourth group, essays 7 to 10, where the question is raised of whether deliberation is a causal, deterministic process, or a paradigmatic example of a rational, controlled, but indeterministic process.

In the fifth group, essay 11 examines the claim of Benjamin Libet that a neural "readiness potential" in the brain precedes all conscious decisions, with the consequence that action initiation would not be under conscious control. The essay presents arguments against Libet's hypothesis.

The sixth group, essays 12 to 14, deals with the question of whether regarding the world as consisting of three-dimensional objects moving in time, or alternatively of static four-dimensional objects extended in time, are metaphysically equivalent positions. The seventh group, essays 15 to 18, introduces the notion of branching spacetime, and provides a philosophical account of laws of nature in terms of branching. In a branching world, the flow of time consists of the progressive random selection of a single "actual" branch, out of a set of possible ones, and the vanishing of the rest. The process corresponds to what underlies the "collapse of the wave-function" in quantum mechanics, and branching spacetime provides an explanation of the mysterious "non-local influences" associated with entangled two-particle quantum states.

Finally the last group, essays 19 to 21, comprises three separate themes. The first is a discussion of the "hierarchical" structure of science, focussing on the place occupied by biology. The second essay offers a solution to the age-old problem of reconciling human free will with God's omniscience, and the third and closing essay presents what appears to be a totally insoluble philosophical problem.

2. Hilbert, Gödel, and the Consistency of Arithmetic

In 1900, David Hilbert presented a list of 23 unsolved mathematical problems to the Paris conference of the International Congress of Mathematicians. The second problem on this list was to prove the consistency of the arithmetical axioms, i.e. to show that from the axioms no pair of theorems |- A, and |- not-A, were both derivable. In Hilbert's words:

> But above all I wish to designate the following as the most important among the numerous questions which can be asked with regard to the axioms: *To prove that they are not contradictory, that is, that a definite number of logical steps based upon them can never lead to contradictory results.*

From 1900 to 2011 Hilbert's second problem has remained unresolved, and his program of proving arithmetical consistency unachieved. The conventional wisdom regarding it is that in 1931 Kurt Gödel showed that it was unachievable, i.e. that even if arithmetic were consistent, this fact could never be proved. Gödel's proof falls into the same category as another well-known impossibility proof: the impossibility of deriving Euclid's parallel postulate from the other axioms of geometry.

Briefly, Gödel devised a method of assigning numerical codes to arithmetical expressions like $2 + 3 = 5$, and $(x)(\exists y)(y = x + 1)$ (i.e., for every number x there exists a y such that y is one unit greater than x). The codes for such formulae are known as their *Gödel numbers* (or, for short, GN). Gödel also assigned codes to *sequences* of expressions. Thus every sequence of assertions of PA, the formal axiomatization of arithmetic known as "Peano arithmetic", was assigned a GN. If such a sequence had the property that each member of the sequence was either a PA axiom, or followed from earlier members by one of PA's rules of inference, then its GN was the GN of a *proof* in PA, and the GN of the last member of the sequence was the GN of a PA *theorem*. Let "$\ldots x \ldots y \ldots$." be the statement that "The sequence of formulae with Gödel number x is a proof in PA of the formula with GN y".

Next, Gödel formalized the concept of *substitution* by defining "sub(y,z)" to be the Gödel number of the formula that results from substituting y for all occurrences of free variables in the formula with Gödel number z. Consider

$$\sim(\exists x)(\ldots x \ldots \text{sub}(y,z) \ldots), \tag{1}$$

which states that there is no proof of the formula that results from substituting y for all occurrences of free variables in the formula with GN z. Let the Gödel number of (1) be g. Since y and z occur free in (1),

$$\sim(\exists x)(\ldots x\ldots \mathrm{sub}(g,g)\ldots) \tag{2}$$

is the PA formula that results from substituting g for all free variables in the formula whose Gödel number is g. Let G abbreviate formula (2). In a perfectly precise way, G says of itself that it is unprovable, and indeed Gödel showed that neither G nor its negation ~G is a PA theorem. In Gödel's words, G is a "formally undecidable" expression, and its existence shows that mathematics is and always will be an incomplete science.

The connection with consistency is as follows. Because of the theorem $\vdash (p \,\&\, {\sim}p) \supset q$ of propositional logic, an inconsistent system in which both A and its negation ~A are theorems is also a system in which *every* well-formed expression or proposition is provable. Consequently a system S is consistent if and only if there exists at least one unprovable S-proposition. Let "Dem(x,y)" abbreviate "The sequence of formulae with Gödel number x is a proof in PA of the formula with GN y", i.e.…x…y… . Then to assert that PA is consistent is to assert $(\exists y)(x) \sim \mathrm{Dem}(x,y)$, i.e. "there is a y such that for all x, the sequence of formulae with GN x is not a proof of the formula with GN y". Let "Cons" be the abbreviation for "$(\exists y)(x) \sim \mathrm{Dem}(x,y)$". Then Gödel was able to prove $\vdash$ Cons $\supset$ G as a PA theorem. If $\vdash$ Cons were also provable, the rule of detachment would yield $\vdash$ G, contrary to Gödel's previous proof that G was not a PA theorem. It follows that Cons is not a PA theorem, i.e. that the consistency of arithmetic is not provable in PA.

Does this mean that Hilbert's Program, calling for an axiomatic formalization of mathematics together with a proof that the resulting system engenders no contradictions, is doomed to failure? Not necessarily. In essay 1 below is found a different kind of consistency proof of arithmetic, a proof based not on deriving theorems from axioms but on a *semantic model* that includes a non-linguistic domain of objects. This proof satisfies Hilbert's requirement of *finitism*, namely that the overall argument, and each of the sub-proofs on which it is based, should consist of a finite number of steps. A key requisite is that every model in the semantics for PA must be based on a finite domain. The conclusion is, that despite Gödel's incompleteness theorems of 1931, the consistency of arithmetic is provable. The proof requires that PA's mathematical axioms, symbols and syntax be supplemented by a non-linguistic, semantic domain of objects.

3. Provability and Truth

Essays 2 and 3 concern the question of whether a computational device such as a Turing machine is capable of differentiating between what is true and what is provable. As was seen above, Gödel constructs the formula G:

$$\text{G.}\quad \sim(\exists x)(\ldots x\ldots \mathrm{sub}(g,g)\ldots),$$

which says of itself that it is unprovable. Gödel goes on to show that, on the assumption that PA is consistent, G is not a theorem of PA. Therefore what G says is true. On the other hand, if PA should turn out to be inconsistent, G would turn out to be false, since every well-formed formula is provable in an inconsistent system. Consequently G is either unprovable and true, or provable and false. Either way, truth and provability part company, and essays 2 and 3 below deal with this theme.

4. Deliberation, Decision and Free Will

The third group, essays 4, 5 and 6, deals with practical deliberation, Aristotle's *bouleusis*. This concerns our ability, as agents, consciously to choose which one of a set of different possible alternative actions to perform. The process of deliberation comprises three successive stages:

1. Listing the set of possible alternative actions open at the time of the deliberation.
2. Evaluating these options.
3. Deciding which one to perform.

The question is raised, whether and in what sense we are free to choose and perform whatever option we wish, or whether our choices are causally determined by factors of which we are unaware. Is decision-making a deterministic or an indeterministic process? Is it a matter of pure luck which of the available options we choose? Essay 6 argues that although deliberation is indeterministic, coming to a decision is not a matter of luck at all, but a reasoned process that can in principle be explained to a friend or a third party.

5. Rational, Controlled, Indeterministic Processes

The theme of the previous section is continued in the fourth group, essays 7 to 10, which raise the question of whether deliberation is a process based on causes, or on reasons. The so-called "causal theory of action" maintains that decision-making is a causal process, despite the fact that when we report the results of a deliberation we normally speak of the *reasons for*, rather than the *causes of*, any given decision. Although not composed of causally linked stages, the steps in a deliberative process are tightly and consciously controlled by the deliberator, and *bouleusis* belongs to a broad family of rational, controlled, indeterministic (RCI) processes such as walking down a crowded sidewalk without colliding with other people, playing the violin, sinking long putts in golf, and playing chess. When we deliberate, we necessarily regard the future as being "open", in the sense that all options remain choosable, and on the table, until the final moment of decision. Throughout evolutionary history, human beings have exploited this open-ness, and prospered at the expense of their single-track, single-option competitors.

6. Are Decisions Conscious?

In the fifth group, essay 11 examines the claim of Benjamin Libet that a neural "readiness potential" in the brain precedes all conscious decisions, with the consequence that initiation of decisions and actions would not be under conscious control. The essay presents arguments against Libet's hypothesis, citing examples such as the split-second decisions made by sprinters at the start of the 100-metre dash. The well-known phenomenon of "false starts", dreaded by championship racers, counts against Libet's theory.

7. 3D/4D Equivalence, Branching Spacetime, Laws of Nature, Time Flow, and Quantum Mechanics

Does the world we live in contain three-dimensional objects that endure and move about in time, or does it consist of fixed four-dimensional objects with temporal extension but no motion? In essays 12–14 it is contended that the 3D and the 4D descriptions of the world are equivalent, and that it is not a question of one being true and the other false. In everyday life, 3D descriptions are appropriate and perfectly adequate. But in other contexts, understanding requires a 4D framework. The twins paradox, for example, and the train-tunnel paradox, cannot be explained and understood in 3D terms (essay 12). On the other hand, the nature of a clock is essentially that of a three-dimensional object functioning in time: no four-dimensional description seems adequate. And when assigning a metric to 4D spacetime, a clock is needed to define the relationship of temporal equality of intervals (essay 13). In essay 15, the concept of branching 4D spacetime is introduced, and laws of nature are analyzed as patterns of events and states of affairs instantiated in the branches. Branching spacetime structures are dynamic, not static, and the flow of time corresponds to *branch attrition*. Branch attrition consists of the progressive on-going random selection of a single "actual" branch out of the multiplicity of physically possible branches that split off from each other at each "now" moment, and the vanishing of the rest (essays 16 and 17). This "transition from potentiality to actuality" corresponds in certain circumstances to the concept of "measurement", or "collapse of the wave-function", in quantum mechanics, and in essay 18 the branch points or "surfaces" along which spacetime branches split are identified as spacelike hypersurfaces stretching from one end of the universe to the other.

Branch selection along infinite hyperplanes provides an explanation of the long-distance correlations observed in the EPR experiment. If the left particle in a two-particle entangled quantum state is measured, it will yield each one of the outcomes "spin-up" and "spin-down" on average 50% of the time. But if the right particle is measured "spin-up", the left particle will *always* be "spin-down", even though it may be far removed from its twin. If the right particle is "spin-down", the left will always be "spin-up". How does the left particle "know" what its twin has been measured, given that it may be travelling away from it at the speed of light? Essays 15 and 18 furnish branching spacetime diagrams to show how branch attrition along spacelike hyperplanes can have the effect of instantaneously conveying information of this kind from one member of an entangled state to its twin.

8. Downward Causation and Biological Information

Essay 19 introduces the Oppenheim-Putnam conception of the unity of science as a "layered" structure, with elementary particle physics at the bottom, and solid state physics, chemistry, biology, psychology, etc. above. At each level, the laws of the lower level continue to apply at the level above, plus other laws particular to the higher level. "Upward causation" means that a process governed by the laws of level X may produce effects at level X + 1, while "downward causation" implies that in exceptional cases a process at level X + 1 may influence events at level X. For example, the structure and functioning of the DNA molecule at the chemical level ("hardware"), generates developmental information ("software") at the higher level of biology ("upward causation"), and this information in turn guides protein synthesis of the right kind and at the right location for an organism's physical growth and development ("downward causation"). Essay 19 argues that in addition to the digital information encoded and stored in DNA, growth and development of an organism may also be guided by a 4D "morphogenetic" or "analog information field", similar to the 4D gravitational field that guides the motion of 3D bodies in general relativity. In general relativity, "matter tells spacetime how to curve, and spacetime tells matter how to move". In single-celled protozoans such as *Stentor coerulus*, which grow and regenerate without cell division, the information required for such regeneration would not be digital information from DNA, but analog "pattern information" stored in a morphogenetic field.

9. God's Omniscience and Human Free Choice

For centuries it has been debated whether God's ability to foreknow the future is consistent with human free choice. In essay 20 it is suggested that if truth is a "supervenient" quality, a new argument reconciling omniscience and free will can be devised. Consider the difference between (i) the world of events, existing in space and time, and (ii) empirical propositions aimed at describing these events, which may be either true or false. Analogous to the way in which mental events M have been claimed to "supervene" upon physical events P in the brain, such that M depends upon P, and every difference in M logically implies a difference in P, so it is argued in essay 20 that the truth of true empirical propositions "supervenes" upon events. In a sense, such propositions "float above" the world of events, and depend upon these events, while, conversely, the world of events neither floats above nor depends upon true propositions. For example, "the sky is blue" is true *because* the sky is blue, but the reason why the sky is blue is not because the proposition "the sky is blue" is true. Suppose now that X freely decides, after due deliberation, to take the bus rather than walking home tonight. If God is omniscient, he knows the truth of all true propositions, and he knows from all eternity that X will take the bus tonight. Can X refrain from taking the bus? Not if doing so would cause something Y to be false, when God knows Y to be true. It would seem that X cannot choose freely what to do, if God already knows what he will do. But wait. If what God knows depends on what is true, and if what it true supervenes and depends on what happens in the world of events, then God's foreknowledge can depend upon what X freely chooses to do. Freewill is compatible with God's omniscience.

10. An Insoluble Problem

A famous but unscrupulous 20th century artist meets a 21st century critic who travels backward in time carrying a portfolio of reproductions of the artist's works (essay 21). When the critic departs he unfortunately leaves the portfolio behind, and the artist spends the rest of his life copying the reproductions. The copies form the basis of his future fame. Problem: what is the source of the artistic originality and creativity that the copies manifestly exhibit? The author offers a prize for its solution.

{1}

The Consistency of Arithmetic

God exists because mathematics is consistent, and the Devil exists because we can't prove it.

—André Weil

Is Peano arithmetic (PA) consistent? This paper contains a proof that it is: a proof, moreover, that does not lie in deducing its consistency as a theorem in a system with axioms and rules of inference. Gödel's second incompleteness theorem states that, if PA is consistent, its consistency cannot be proved in PA. But to deduce its consistency in some stronger system PA+ that includes PA is self-defeating, since if PA+ is itself inconsistent, the proof of PA's consistency is worthless. In an inconsistent system everything is deducible, and therefore nothing is provable. If there is to be a genuine proof of PA's consistency, it cannot be a proof *relative to* the consistency of some other, stronger system, but an absolute proof, such as the proof of consistency of two-valued propositional logic using truth tables. Axiomatic proofs we may categorize as "syntactic", meaning that they concern only symbols and the derivation of one string of symbols from another, according to set rules. "Semantic" proofs, on the other hand, differ from syntactic proofs in being based not only on symbols but on a non-symbolic, non-linguistic component, a *domain of objects*. If the sole paradigm of "proof" in mathematics is "axiomatic proof", in which to prove a formula means to deduce it from axioms using specified rules of inference, then Gödel indeed appears to have had the last word on the question of PA consistency. But in addition to axiomatic proofs there is another kind of proof. In this chapter I give a proof of PA's consistency based on a formal semantics for PA. To my knowledge, no semantic consistency proof of Peano arithmetic has yet been constructed.

The difference between "semantic" and "syntactic" theories is described by van Fraassen in his book *The Scientific Image*:

> The syntactic picture of a theory identifies it with a body of theorems, stated in one particular language chosen for the expression of that theory. This should be contrasted with the alternative of presenting a theory in

> the first instance by identifying a class of structures as its models. In this second, semantic, approach the language used to express the theory is neither basic nor unique; the same class of structures could well be described in radically different ways, each with its own limitations. The models occupy centre stage. (1980, p. 44)

Van Fraassen gives the example on p. 42 of a consistency proof in formal geometry that is based on a non-linguistic model. Suppose we wish to prove the consistency of the following geometric axioms:

A1. For any two lines, there is at most one point that lies on both.
A2. For any two points, there is exactly one line that lies on both.
A3. On every line there lie at least two points.

The following diagram shows the axioms to be consistent:

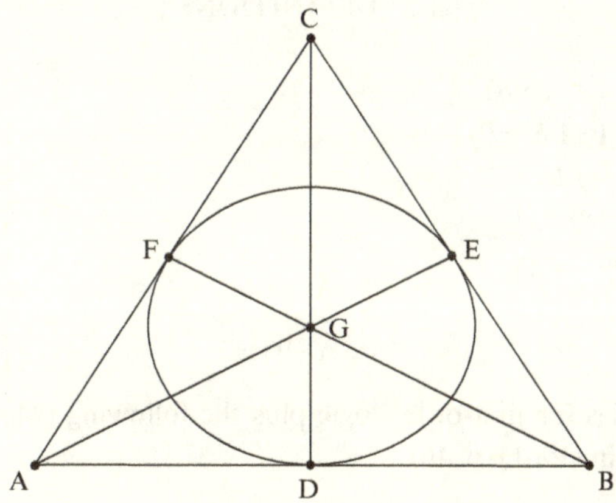

FIGURE 1

The consistency proof is not a "syntactic" one, in which the consistency of A1–A3 is derived as a theorem of a deductive system, but is based on a non-linguistic structure. It is a semantic, as opposed to a syntactic, proof. The proof constructed in this chapter, like van Fraassen's, is based on a non-linguistic component, not a diagram in this case but a physical domain of three-dimensional cube-shaped blocks. In Section 1, Peano arithmetic is presented as an axiomatic system, and in Section 2, formal semantics based on block domains are laid down for it.

1. The System PA

1.1. PRIMITIVE SYMBOLS

Logical symbols: &, ~, =, ∀
Arithmetic symbols: 0, S, +, ×
Variables: *x*, *y*, *z*, . . .

1.2. RULES OF FORMATION

1.2.1 For Terms:

(i) All variables, plus 0, are terms.
(ii) If X and Y are terms, then so are SX, $X + Y$, and $X \times Y$.
(iii) Nothing else is a term.

1.2.2 For wffs:

(i) If X and Y are terms, then $X = Y$ is an atomic wff.
(ii) If A is a wff and x is a variable, then $(\forall x)A$ is a wff.
(iii) If A and B are wffs, then so are $\sim A$ and A & B.
(iv) Nothing else is a wff.

1.3. DEFINITIONS

$$A \supset B =_{df} \sim(A \,\&\, \sim B)$$
$$A \vee B =_{df} \sim(\sim A \,\&\, \sim B)$$
$$(x)A =_{df} (\forall x)A$$
$$(\exists x)A =_{df} \sim(x)\sim A$$
$$\mathbf{1} =_{df} \mathrm{S}0$$

1.4. AXIOMS

Axioms and rules for first-order logic plus the following (Mendelson (1964), p. 103; Goodstein (1961), p. 46):

A1. $(x = y) \supset (x = z \supset y = z)$
A2. $\sim(\mathrm{S}x = 0)$
A3. $(x = y) \supset (\mathrm{S}x = \mathrm{S}y)$
A4. $(\mathrm{S}x = \mathrm{S}y) \supset (x = y)$
A5. $x + 0 = x$
A6. $x + \mathrm{S}y = \mathrm{S}(x + y)$
A7. $x \times 0 = 0$
A8. $x \times \mathrm{S}y = (x \times y) + x$
Induction rule: From |- $F(0)$ and |- $(x)(Fx \supset F(\mathrm{S}x))$ infer |- $(x)Fx$.

2. Formal Semantics for PA

The semantics presented in this chapter I call "block semantics", for reasons that will become clear.[1] Block semantics is based on domains consisting of

[1] For helpful suggestions and advice in developing block semantics I am indebted to several colleagues.

cube-shaped objects of the same size, e.g., children's wooden building blocks. These can be arranged either in a *linear* array or in a *rectangular* array, i.e., either in a row with no space between the blocks or in a rectangle composed of rows and columns. A linear array can consist of a single block, and the order of individual blocks in a linear or rectangular array is irrelevant. Given three blocks A, B and C, the linear arrays ABC and BCA are indistinguishable. Two linear arrays can be joined together, or *concatenated*, into a single linear array, and a rectangle can be rearranged or transformed into a linear array by successive concatenation of its rows. The result is called the "linear transformation" of the rectangle. An essential characteristic of block semantics is that every domain of every block model is *finite*. In this respect it differs from Tarski's semantics for first-order logic, which permits infinite domains. But although every block model is finite, there is no upper limit to the number of such models or to the size of their domains.

It should be emphasized that block models are physical models, the elements of which can be physically manipulated. Their manipulation differs in obvious and fundamental ways from the manipulation of symbols in formal axiomatic systems and in mathematics. For example, the transformations just described, in which two linear arrays are joined together to form one array, or a rectangle of blocks is re-assembled into a linear array, are physical transformations, not symbolic transformations.

(It has been suggested to me by some colleagues that, in the interests of constructing a *fully formalized* proof, an "ersatz" facsimile of block models could be made out of linguistic items rather than three-dimensional blocks, for example, domains consisting entirely of tokens of the letter "*a*". There could be linear sequences of *a*'s and rectangular arrays of *a*'s, somewhat in the style of "ersatz" possible worlds in modal semantics consisting of maximal consistent sets of propositions (see, e.g., Lewis (1986), pp. 136–144). This is true I suppose, but the proposal does not convert the proposed consistency proof for arithmetic into a fully formalized *axiomatic* proof any more than a consistency or independence proof in modal logic based on ersatz possible worlds becomes an axiomatic proof rather than a *semantic* one.)

A *semantic block model* M consists of a domain D and an assignment function v. M = <D, v>. A *domain* D is a set of blocks, and the *assignment function* v connects terms of PA with linear or rectangular arrays of blocks. Variables x, y, z are each assigned linear arrays of blocks by the function v, and 0 is assigned the "empty", or null, array Ø. The term $x + y$ is assigned the linear concatenation of the arrays assigned to x and y, and $x \times y$ is assigned the rectangular array with sides $v(x)$ and $v(y)$. By a *linear transformation*, a rectangular array can be reassembled into a linear array formed by successively concatenating the rows of the rectangle. The term Sx is assigned the linear transformation of whatever is assigned to x, plus one block more. A formal definition of the function v is given in upcoming Section 4. But first a problem must be addressed concerning assignments to terms such as SSSx, or SSx × SSx, in models with domains of, say, only two blocks.

In a two-block model, assignments can be made to 0, x, Sx, and SSx, but not to SSSx. In such a model, the value of v(SSSx) is undefined, and v is a partial function, not a total function. This complicates the semantics somewhat, since the valuation function v_M over a model M, which takes formulae into truth-values, will also, like v, be a partial function. When M's domain contains only two blocks, v_M(x = SSS0), and v_M(S0 + SS0 = SSS0) cannot take the value "true", or the value "false", but can only be "undefined". Consequently not all theorems of PA are true in all models, but only in "sufficiently large" models. It is possible to construct a rigorous formal semantics in which the assignment and valuation functions are partial functions, and the value "U" (or "undefined") takes its place beside the regular truth-values T and F. But as will be seen, there is a simpler solution that results in a bivalent rather than a trivalent semantics. This will involve constructing a revamped axiomatic system for PA, based not on functional expressions like "x + y", "x × y" and "Sx", but on relational predicates like Sxyz ("the sum of x and y is z") and Pxyz ("the product of x and y is z"). Such an approach, reminiscent of Russell's theorem on the eliminability of definite descriptions, is found in Abraham Robinson (1965a, 1965b).[2] As it turns out, the semantics for PA using Robinson axioms rather than the traditional axioms of section 1 above are bivalent, meaning that every wff in every model is either true or false. The Robinson-style axiomatic system is marginally more complex than the usual one, but this is more than compensated for by the simplicity of its semantics.

3. PA in Traditional Form vs. the Robinson-Style System RPA

The usual axiomatic basis for the system PA is given in Section 1. When function symbols are replaced by relations, the resultant Robinson-style system RPA is as follows.

3.1. PRIMITIVE SYMBOLS

(i) Logical symbols: &, ~, =, ∀
(ii) Variables: x, y, z,...
(iii) Constants: 0, 1
(iv) Three-place relations Sxyz ("The sum of x and y is z") and Pxyz ("The product of x and y is z").

Variables and constants are *terms*.

[2] I owe the reference to Robinson's relational arithmetic to Alasdair Urquhart.

3.2. RULES OF FORMATION

(i) Where a, b, and c are terms, $a = b$, Sabc and Pabc are wffs.
(ii) If A is a wff and x is a variable, then $(\forall x)A$ is a wff.
(iii) If A and B are wffs, so are $\sim A$ and $A \ \& \ B$.
(iv) Nothing else is a wff.

3.3. DEFINITIONS

Df I. $A \supset B$ is defined as $\sim(A \ \& \sim B)$.
Df II. $(x)A$ is defined as $(\forall x)A$.
Df III. $(\exists x)A$ is defined as $\sim(x)\sim A$.
Df IV. Cxy is defined as S$x1y$.
(Cxy is read "The successor of x is y").

Additional definitions are needed for the derivation of the axioms of PA in RPA. These definitions are given in Section 8.

The axioms and rules of inference for RPA are those of first-order logic with identity, plus the following:

R1. S$x0x$
R2. ~S$x10$
R3. (Sxyz & Swyz) $\supset x = w$
R4. P$x00$
R5. P$x1x$
R6. $(x)(y)(z)(\exists s)(\exists t)(\exists u)$(S$yzs$ & Sxsu & Sxyt & Stzu)
R7. $(x)(y)(z)(\exists r)(\exists s)(\exists t)(\exists u)$(S$yzr$ & Pxrs & Pxyt & Pxzu & Stus)
R8. $(\exists y)$[Cxy & (z)(C$xz \supset z = y$)]
R9. $(\exists z)$[Sxyz & (w)(S$xyw \supset w = z$)]
R10. $(\exists z)$[Pxyz & (w)(P$xyw \supset w = z$)]

(R8–R10: Existence and uniqueness axioms for "successor", "sum" and "product".)

Induction rule. From |- $F(0)$ and |- $(x)(y)((Fx$ & C$xy) \supset Fy)$ infer |- $(x)Fx$.

Section 8 contains formal derivations of axioms A1–A8 and PA's induction rule from axioms R1–R10, RPA's induction rule, and other definitions stated there. RPA is consequently a complete system of Peano arithmetic.

4. Formal semantics for RPA

As in Section 2 (earlier), semantic models M = <D,v> for RPA consist of a domain D of blocks and an assignment function v that assigns linear or rectangular arrays of blocks to variables, a single block to the constant 1, and the empty, or null, sequence Ø to the constant 0.

(i) Two linear arrays are *equal* if they form a rectangle when placed beside each other,
(ii) A rectangular array and a linear array are equal if the latter is a linear transformation of the former, and
(iii) Two rectangles are equal if their linear transformations are equal.

4.1. VALUATION FUNCTION

The *valuation function* v_M with respect to a model M is a function that takes every wff A of RPA into the set of truth-values {T, F}. It is defined inductively. Except where quantified or where it is explicitly stated to the contrary, the letters $x, y, z, \ldots$ henceforth stand for either variables or constants.

4.1.1 Basis

A is an atomic wff.

(i) $v_M(x = y) = \mathrm{T}$ iff $v(x)$ and $v(y)$ are equal.
(ii) $v_M(\mathrm{C}xy) = \mathrm{T}$ iff the linear transformation of $v(y)$ equals the linear transformation of $v(x)$ plus one block.
(iii) $v_M(\mathrm{S}xyz) = \mathrm{T}$ iff the linear transformation of $v(x)$ concatenated with the linear transformation of $v(y)$ equals the linear transformation of $v(z)$.
(iv) $v_M(\mathrm{P}xyz) = \mathrm{T}$ iff the rectangular array with sides equal to the linear transformations of $v(x)$ and $v(y)$ is equal to $v(z)$.

4.1.2 Induction Step

Assume $v_M(A)$ and $v_M(B)$ have already been defined. Then:

(i) $v_M(\sim A) = \mathrm{T}$ iff $v_M(A) = \mathrm{F}$.
(ii) $v_M(A \mathbin{\&} B) = \mathrm{T}$ iff $v_M(A) = \mathrm{T}$ and $v_M(B) = \mathrm{T}$.
(iii) $v_M((x)A) = \mathrm{T}$ iff for all models M′ = <D′,v′>, where D is a subset of D′ and where v' differs from v at most in assignment to x, $v_{M'}(A) = \mathrm{T}$.
(iv) $v_M((\exists x)A) = \mathrm{T}$ iff there is a model M′ = <D′,v′>, where D is a subset of D′ and where v' differs from v at most in assignment to x, such that $v_{M'}(A) = \mathrm{T}$.

4.2. TRUTH IN A MODEL

A formula A is *true in a model M* iff $v_M(A) = \mathrm{T}$.

4.3. VALIDITY

A is *valid* iff A is true in all non-empty models.

(The restriction of validity to truth in *non-empty* models is necessitated by the constant 1. $0 \neq 1$ is intuitively valid, but it is not true in empty models.)

As stated earlier, the reason for switching from the traditional system PA of Peano arithmetic to the Robinson-style system RPA is that the block semantics for RPA are bivalent, whereas those for PA are not. In models with domains of just two elements, the formula

$$x = \mathrm{SSS0}$$

is neither true nor false but is "undefined" in the semantics for PA, as is the theorem

$$\mathrm{S0} + \mathrm{SS0} = \mathrm{SSS0}.$$

In the block semantics for PA, validity cannot be defined as truth in all non-empty models. But in RPA, which lacks functional expressions like "S" and "+," the formula that corresponds to S0 + SS0 = SSS0 is $(x)(y)(z)[(\mathrm{C0}x\ \&\ \mathrm{C}xy\ \&\ \mathrm{C}yz) \supset \mathrm{S}xyz]$, and this formula is true in one-element and two-element models. To show this we argue by *reductio*:

1. Assume there is a model $\mathrm{M} = <\mathrm{D},v>$ such that $v_{\mathrm{M}}(x)(y)(z)[(\mathrm{C0}x\ \&\ \mathrm{C}xy\ \&\ \mathrm{C}yz) \supset \mathrm{S}xyz] = F$.
2. From 1, $v_{\mathrm{M}}(\exists x)(\exists y)(\exists z) \sim [(\mathrm{C0}x\ \&\ \mathrm{C}xy\ \&\ \mathrm{C}yz) \supset \mathrm{S}xyz] = T$.
3. There is an $\mathrm{M}' = <\mathrm{D}',v'>$, where D is a subset of D′ and v' differs from v at most in assignment to x, such that $v_{\mathrm{M}'}(\exists y)(\exists z) \sim [(\mathrm{C0}x\ \&\ \mathrm{C}xy\ \&\ \mathrm{C}yz) \supset \mathrm{S}xyz] = T$.
4. There is an $\mathrm{M}'' = <\mathrm{D},''v''>$, where v'' differs from v' at most in assignment to y, such that $v_{\mathrm{M}''}(\exists z) \sim [(\mathrm{C0}x\ \&\ \mathrm{C}xy\ \&\ \mathrm{C}yz) \supset \mathrm{S}xyz] = T$.
5. There is an $\mathrm{M}''' = <\mathrm{D}''',v'''>$, where v''' differs from v'' at most in assignment to z, such that $v_{\mathrm{M}'''} \sim [(\mathrm{C0}x\ \&\ \mathrm{C}xy\ \&\ \mathrm{C}yz) \supset \mathrm{S}xyz] = T$.
6. From 5, $v_{\mathrm{M}'''}(\mathrm{C0}x\ \&\ \mathrm{C}xy\ \&\ \mathrm{C}yz) = T$ and $v_{\mathrm{M}'''}(\mathrm{S}xyz) = F$.
7. Assignments to x, y, and z that satisfy $v_{\mathrm{M}'''}(\mathrm{C0}x\ \&\ \mathrm{C}xy\ \&\ \mathrm{C}yz) = T$ are the linear arrays $v'(x) = 1$ block, $v''(y) = 2$ blocks, and $v'''(z) = 3$ blocks. (Recall that $\mathrm{D} \subseteq \mathrm{D}'''$.) If $v_{\mathrm{M}'''}(\mathrm{C0}x\ \&\ \mathrm{C}xy\ \&\ \mathrm{C}yz) = T$, then $v_{\mathrm{M}'''}(\mathrm{S}xyz) = T$ also, which contradicts line 6. Consequently the assumption on line 1 is false, and there is no model M in which $(x)(y)(z)[(\mathrm{C0}x\ \&\ \mathrm{C}xy\ \&\ \mathrm{C}yz) \supset \mathrm{S}xyz]$ takes the value F.

The net result of moving from a formalized Peano arithmetic containing term-forming functional expressions to an arithmetic containing only sentence-forming relational expressions is that the semantics for the latter are bivalent, whereas the semantics for the former require that the truth-values of some formulae, in some models, be "undefined". Although Robinson is the first (and to the author's knowledge the only) logician to propose a relation-based formal arithmetic, it is unlikely (but not impossible) that his motivation was to achieve a bivalent semantics. We return to a general discussion of Robinson's system in Section 7.

5. Semantic Proof of the Consistency of RPA

Given the definition of "validity" of a formula A in the preceding section, the way to proving RPA consistent is open. If it can be shown that all RPA axioms are valid, and that the rules of inference preserve validity, then all the theorems of RPA will be valid. But it is impossible for two formulae A and ~A to both be valid: If one takes the value T, the other takes the value F. Hence RPA is a consistent system. Also, if RPA is consistent, then PA is consistent too, since, as is shown in Section 8, all PA theorems are derivable in RPA. It follows that if all RPA theorems can be shown to be valid, then Peano arithmetic will be consistent.

We argue case by case that the axioms of RPA in Section 3 are valid. The arguments are generally by *reductio.*

AXIOM R1.

Assume for *reductio* that there is a model M = <D,*v*> in which R1 is false.

1. $v_M(Sx0x) = F$ — Assumption
2. From line 1, the concatenation of the linear transformation of $v(x)$ with $v(0)$, the null array, is not equal to $v(x)$. But this is impossible. Consequently the assumption at line 1 is false, and R1 is true in all models, i.e. valid.

AXIOM R2.

1. $v_M({\sim}Sx10) = F$ — Assumption
2. $v_M(Sx10) = T$ — 1
3. From line 2, the concatenation of a linear array $v(x)$ with a single additional block equals the null array. But this is absurd. Hence line 1 is false.

AXIOM R3.

1. $v_M[(Sxyz \,\&\, Swyz) \supset x = w] = F$ — Assumption
2. $v_M(Sxyz \,\&\, Swyz) = T$ and $v_M(x = w) = F$ — 1

3. From line 2, the concatenation of $v(x)$ with $v(y)$ equals $v(z)$, and the concatenation of $v(w)$ with $v(y)$ equals $v(z)$, but $v(x)$ and $v(w)$ are not equal. This is impossible, hence line 1 is false.

AXIOM R4.

1. $v_M(\text{P}x00) = \text{F}$ Assumption
2. Any rectangle, one of the sides of which is the null array, is null. Hence in any model M, $v_M(\text{P}x00)$ is true, and line 1 is false.

AXIOM R5.

1. $v_M(\text{P}x1x) = \text{F}$ Assumption
2. A rectangle with a unit side is identical with the linear array consisting of the other side. Hence $v_M(\text{P}x1x)$ is true in all models.

AXIOM R6.

1. $v_M[(x)(y)(z)(\exists s)(\exists t)(\exists u)(\text{S}yzs \;\&\; \text{S}xsu \;\&\; \text{S}xyt \;\&\; \text{S}tzu)] = \text{F}$ Assumption
2. From 1, $v_M[(\exists x)(\exists y)(\exists z)(s)(t)(u) \sim (\text{S}yzs \;\&\; \text{S}xsu \;\&\; \text{S}xyt \;\&\; \text{S}tzu)] = \text{T}$
3. From 2, there is a model $\text{M}' = \langle \text{D}', v' \rangle$, where D is a subset of D′, such that
$$v_{M'}[(s)(t)(u) \sim (\text{S}yzs \;\&\; \text{S}xsu \;\&\; \text{S}xyt \;\&\; \text{S}tzu)] = \text{T}$$
4. From 3, for all models M″, where D′ is a subset of D″, and v'' differs from v' at most in assignment to *s, t* and *u*,
$$v_{M''}[\sim (\text{S}yzs \;\&\; \text{S}xsu \;\&\; \text{S}xyt \;\&\; \text{S}tzu)] = \text{T},$$
i.e.,
$$v_M(\text{S}yzs \;\&\; \text{S}xsu \;\&\; \text{S}xyt \;\&\; \text{S}tzu) = \text{F}.$$
5. In line 4,
$v_{M''}(\text{S}yzs) = \text{T}$ iff the concatenation of $v''(y)$ and $v''(z)$ is $v''(s)$,
$v_{M''}(\text{S}xsu) = \text{T}$ iff the concatenation of $v''(x)$ and $v''(s)$ is $v''(u)$,
$v_{M''}(\text{S}xyt) = \text{T}$ iff the concatenation of $v''(x)$ and $v''(y)$ is $v''(t)$, and
$v_{M''}(\text{S}tzu) = \text{T}$ iff the concatenation of $v''(t)$ and $v''(z)$ is $v''(u)$.
6. Among all the models M″ of line 4, $v'(x)$, $v'(y)$ and $v'(z)$ are already specified by M′, and consequently $v''(x)$, $v''(y)$ and $v''(z)$ are also specified (since v'' differs from v' at most in assignment to *s, t* and *u*). We now select a particular model M″ in which we choose $v''(s)$ to denote the concatenation of the array $v''(y)$ with the array $v''(z)$, $v''(t)$ to denote the concatenation of $v''(x)$ and $v''(y)$, and $v''(u)$ to denote the concatenation of $v''(x)$ and $v''(s)$. That is to say, $v''(u)$ denotes:
The concatenation of $v''(x)$ with (the concatenation of $v''(y)$ with $v''(z)$).
But since in all block models concatenation is associative, i.e.,
x CONCAT (y CONCAT z) = (x CONCAT y) CONCAT z,
where *x, y* and *z* are linear arrays of blocks, $v''(u)$ will also denote:
(the concatenation of $v''(x)$ with $v''(y)$), concatenated with $v''(z)$,

i.e.,
$v''(t)$ concatenated with $v''(z)$.
And if $v''(u)$ denotes the concatenation of $v''(t)$ and $v''(z)$ in M″, then
$v_{M''}(Stzu) = T$.

7. A particular M″ has been selected in which $v_{M''}(Syzs) = T$, $v_{M''}(Sxsu) = T$, and $v_{M''}(Sxyt) = T$. But in that model, we also necessarily have $v_{M''}(Stzu) = T$, through associativity of concatenation. This contradicts line 4. Consequently, the assumption on line 1 is false.

(Note. The above argument for the validity of Axiom R6 in block semantics illustrates an important difference between block models and the so-called "standard model" of arithmetic. The standard model consists of the natural numbers, including 0, with the operations of addition and multiplication defined on them. If we were to attempt to show that the axiom of associativity for addition, $x + (y + z) = (x + y) + z$, is satisfied in the standard model, we could argue that the truth of the axiom is based on the property of associativity of the operation of addition. But such an argument would beg the question. We can't demonstrate the truth of the axiom by appealing to the associative property of numerical addition that the axiom states, since to deny the truth of the axiom is simply to deny the existence of the property. But physical concatenation of linear arrays of blocks is associative quite independently of whether addition of numbers is associative, and in block semantics the latter (mathematical) property is based on the former (physical) property.)

AXIOM R7.

1. Assume $v_M(x)(y)(z)(\exists r)(\exists s)(\exists t)(\exists u)(Syzr \;\&\; Pxrs \;\&\; Pxyt \;\&\; Pxzu \;\&\; Stus) = F$. In the way that the validity of Axiom R6 is based on the associativity of concatenation, the validity of R7 will follow from the analogue in block semantics of the distributivity of addition over multiplication, $a(b + c) = ab + ac$.
2. From 1, $v_M[(\exists x)(\exists y)(\exists z)(r)(s)(t)(u) \sim (Syzr \;\&\; Pxrs \;\&\; Pxyt \;\&\; Pxzu \;\&\; Stus)] = T$
3. From 2, there is a model M′ = <D′,v′>, where D is a subset of D′, such that
$$v_{M'}[(r)(s)(t)(u) \sim (Syzr \;\&\; Pxrs \;\&\; Pxyt \;\&\; Pxzu \;\&\; Stus)] = T$$
4. From 3, for all models M″, where D″ is a subset of D′, and v'' differs from v' at most in assignment to r, s, t and u,
$$v_{M''}[\sim (Syzr \;\&\; Pxrs \;\&\; Pxyt \;\&\; Pxzu \;\&\; Stus)] = T,$$
i.e.,
$$v_{M''}(Syzr \;\&\; Pxrs \;\&\; Pxyt \;\&\; Pxzu \;\&\; Stus) = F.$$
5. Among the models M″ of line 4, $v''(x)$, $v''(y)$ and $v''(z)$ are already specified by the assignments to x, y and z in M′. We now select a

particular M″ by choosing $v''(r)$ to denote the concatenation of $v''(y)$ and $v''(z)$, $v''(s)$ to denote the rectangle of blocks with sides $v''(x)$ and $v''(r)$, $v''(t)$ to denote the rectangle with sides $v''(x)$ and $v''(y)$, and $v''(u)$ to denote the rectangle with sides $v''(x)$ and $v''(z)$.

6. For simplicity's sake in what follows, the letters *x, y, z, r, s, t* and u will be used in place of their assignment values $v'(x)$, $v'(y)$, $v'(z)$, $v''(r)$, $v''(s)$, $v''(t)$ and $v''(u)$. As in the case of axiom R6, we argue that if $v_{M''}(Syzr)$, $v_{M''}(Pxrs)$, $v_{M''}(Pxyt)$ and $v_{M''}(Pxzu)$ are all true, then so is $v_{M''}(Stus)$, contradicting line 4.
7. In line 4,
   $v_{M''}(Syzr)$ = T iff the concatenation of *y* and *z* is *r*.
   $v_{M''}(Pxrs)$ = T iff the rectangle with sides *x* and *r* is *s*.
   $v_{M''}(Pxyt)$ = T iff the rectangle with sides *x* and *y* is *t*.
   $v_{M''}(Pxzu)$ = T iff the rectangle with sides *x* and *z* is *u*.
8. In block semantics, when we physically bring the two rectangles *t* and *u* together along the side *x* that they have in common, the result is a larger rectangle with sides *x* and $(y + z)$, where $y + z$ is the concatenation of *y* and *z*. From line 7, $(y + z) = r$. Let t' be the linear transformation of the rectangle *t*, let u' be the linear transformation of the rectangle *u*, and let s' be the linear transformation of the rectangle *s*. By the definition of "equality" at the beginning of section 4 above, $t' = t$, $u' = u$, and $s' = s$.
9. Since the large rectangle with sides *x* and *r* is *s*, and since *s* consists of the two smaller rectangles *t* and *u* put together along their common side *x*, we have that $s = t + u$, or equivalently $s' = t'$CONCAT u'. Since $t = t'$, $u = u'$, and $s = s'$, the "concatenation" of the rectangles *t* and *u* is *s*. (This is the first, and only, time that the notion of "concatenating" two rectangles with a common side in block semantics will be appealed to. Such concatenation is a physical operation performed on physical blocks.) But if the concatenation of *t* and *u* is *s*, then $v_{M''}(Stus)$ = T. Consequently if, in the particular model M″ described in line 5, the four statements $v_{M''}(Syzr)$ = T, $v_{M''}(Pxrs)$ = T, $v_{M''}(Pxyt)$ = T and $v_{M''}(Pxzu)$ = T all hold, then $v_{M''}(Stus)$ = T also necessarily holds. In that model, $v_{M''}[(Syzr \,\&\, Pxrs \,\&\, Pxyt \,\&\, Pxzu \,\&\, Stus)]$ = T, contrary to line 4. By *reductio*, the assumption on line 1 is false, and axiom R7 is valid.

AXIOMS R8–R10.

R8. $(\exists y)[Cxy \,\&\, (z)(Cxz \supset z = y)]$	Existence and uniqueness
R9. $(\exists z)[Sxyz \,\&\, (w)(Sxyw \supset w = z)]$	axioms for "successor", "sum" and
R10. $(\exists z)[Pxyz \,\&\, (w)(Pxyw \supset w = z)]$	"product".

There is no difficulty in seeing that these three axioms are valid in block semantics. Suppose for example, for *reductio*, that there is an M where $v_M(\exists y)[Cxy \,\&\, (z)(Cxz \supset z = y)] = F$, i.e. that $v_M(y) \sim [Cxy \,\&\, (z)(Cxz \supset z = y)] = T$. Then for all M′ = <D′,v′>, where D is a subset of D′ and v′ is like v except possibly in assignment to y, $v_{M'} \sim [Cxy \,\&\, (z)(Cxz \supset z = y)] = T$, i.e. $v_{M'}[Cxy \,\&\, (z)(Cxz \supset z = y)] = F$. Select a model M′ in which $v'(y)$ is $v'(x)$ plus one block. Then $v_{M'}(Cxy) = T$, and consequently $v_{M'}[(z)(Cxz \supset z = y)] = F$, i.e. $v_{M'}[(\exists z) \sim (Cxz \supset z = y)] = T$. It follows that there is an M″ = <D″, v″>, where D′ is a subset of D″ and v″ differs from v′ at most in assignment to z, such that $v_{M''}[\sim(Cxz \supset z = y)] = T$, i.e. that $v_{M''}(Cxz) = T$ and $v_{M''}(z = y) = F$. But if in M″ we select $v''(z)$ to be equal to $v'(x)$ plus one block, then $v''(z) = v'(y) = v''(y)$, and $v_{M''}(z = y) = T$. And if in M″ we select $v''(z)$ to be different from $v'(x)$ plus one block, then, since $v''(x) = v'(x)$, $v_{M''}(Cxz) = F$. In either case we get a contradiction, and the hypothesis of the *reductio* is false. So R8 is valid.

The arguments for the validity of axioms R9 and R10 are similar.

In addition to the arithmetical axioms R1–R10, RPA rests upon axiom-schemata for propositional logic that are known to be valid, plus two first-order schemata R11 and R12 and a rule RQ for predicate logic (see, e.g., Church (1956), p. 172). These may be shown to be valid, or in the case of RQ validity-preserving, as follows.

AXIOM-SCHEMA R11.

$(x)(A \supset B) \supset (A \supset (x)B)$, where x does not occur free in A.

1. Assume (for *reductio*) that there is a model M = <D,v> in which $v_M[(x)(A \supset B) \supset (A \supset (x)B)] = F$.
2. $v_M(x)(A \supset B) = T$ and $v_M(A \supset (x)B) = F$ — 1
3. $v_M(A \supset (x)B) = F$ — 2
4. $v_M(A) = T$ and $v_M((x)B) = F$ — 3
5. $v_M(x)(A \supset B) = T$ — 2
6. For all models M′ = <D′,v′>, where D is a subset of D′ and v′ differs from v at most in assignment to x, $v_{M'}(A \supset B) = T$ — 5, 4.1.2 (iii)
7. For each such M′, either $v_{M'}(A) = F$ or $v_{M'}(B) = T$ — 6
8. But since x is not free in A, for each such M′ $v_{M'}(A) = v_M(A)$
9. Therefore, for each such M′, either $v_M(A) = F$ or $v_{M'}(B) = T$ — 7, 8
10. $v_M(A) = T$ — 4
11. Therefore, for each such M′, $v_{M'}(B) = T$ — 9, 10
12. $v_M((x)B) = T$ — 11, 4.1.2 (iii)
13. But $v_M((x)B) = F$ — 4

Contradiction. Hence the assumption on line 1 is false, and R11 is valid.

AXIOM-SCHEMA R12.

$(x)A \supset B$, where B is like A except that B may contain a term *t*, or free occurrences of a variable *y*, wherever A contains free occurrences of *x*.

1. Assume there is a model M = <D,*v*> such that $v_M((x)A \supset B) = F$.
2. $v_M(x)A = T$ and $v_M(B) = F$ 1
3. $v_M(x)A = T$ 2
4. For all M′ = <D′,*v*′>, the assignment functions of which differ from M at most in assignment to *x*, $v_{M'}(A) = T$. 3
5. Since D is a subset of D′, M′ will assign to *y* and the variables of *t* the same linear arrays of blocks that M assigns to *y* and the variables of *t*. (Note that this would not necessarily be the case if D were not a subset of D′.) Hence, for all M′, $v_{M'}(B) = v_M(B)$.
6. $v_{M'}(B) = F$ 2, 5
7. But since B differs from A only in having free *y* or a term *t* wherever A has free *x*, there will be, amongst the models M′, a model M′ = <D′,*v*′>, where D is a subset of D′ and M′ differs from M at most in assignment to *x*, such that $v_{M'}(B) = v_{M'}(A)$. (We simply assign to *y* in M, and consequently in M′, the same array of blocks that M assigns to *x*. Alternatively, let $v(t) = a$ in M, where *a* is some array of blocks. Then, there will be at least one M′, where D is a subset of D′, such that $v'(x) = a$ in M′. In that model M′, $v_{M'}(B) = v_{M'}(A)$.)
8. $v_{M'}(A) = F$ 6, 7, since 6 holds for all M′.

Line 8 contradicts 4. Consequently the assumption on line 1 is false, and R12 is valid.

Next it is shown that the rules of inference of RPA preserve validity.

(i) The rule RQ. From |- A infer |- (*x*)A.
Assume that A is valid but that (*x*)A is not, i.e. $v_M(x)A = F$ for some model M. The latter implies that for some model M′ =<D′,*v*′>, where D is a subset of D′ and *v*′ differs from *v* at most in assignment to *x*, $v_{M'}(A) = F$. But if A is valid, then for all such models M′, $v_{M'}(A) = T$. Hence if A is valid then (*x*)A is valid, i.e. RQ preserves validity.

(ii) Modus ponens being validity-preserving, it remains to consider the rule of induction. This has traditionally been the sticking point in arithmetical consistency proofs, presumably because of the step premiss that states "For all *x* and *y*, if F*x* and if the successor of *x* is *y*, then F*y*". There is no upper bound to the number of iterated steps from F*x* to F*y* before the conclusion (*x*)F*x* is reached, hence the rule of induction is to all appearances non-finitistic. However, if we argue indirectly by *reductio*, the validity-preserving character of the rule can be shown finitistically, by a method describable as "finite descent".

1. Assume that F(0) and $(x)(y)((Fx \,\&\, Cxy) \supset Fy)$ are both valid, but that (*x*)F*x* is not.

2. That is, there is a model M = <D,v> such that $v_M(x)Fx = F$
3. That is, $v_M(\exists x) \sim Fx = T$
4. From 3, there is a model M′ = <D′,v'>, where D is a subset of D′ and M′ differs from M at most in assignment to x, such that $v_{M'}(\sim Fx) = T$
5. In M′, let the predicate ~F be true of N blocks
6. Since $(x)(y)((Fx \,\&\, Cxy) \supset Fy)$ is valid, its antilogism contrapositive $(x)(y)((\sim Fy \,\&\, Cxy) \supset \sim Fx)$ is also valid. Recall that "Cxy" means "the successor of x is y". Therefore, in all models, if the predicate ~F is true of a linear array of N blocks, it is also true of N-1 blocks. And if true of N-1 blocks, it is also true of N-2 blocks. Etc.
7. Applying line 6 repeatedly to line 5, ~F is true in M′ of the empty array Ø.
8. Consequently $v_{M'} \sim F(0) = T$, i.e. $v_{M'}F(0)$ is false, contrary to the assumption that F(0) is valid.
9. By *reductio*, the assumption on line 1 is false, and the rule of induction preserves validity.

This *reductio* argument requires only a finite number of steps, and is consequently finitistic. Note that if we had attempted to argue directly for the validity of induction, i.e. to show that in all models where F(0) and $(x)(y)((Fx \,\&\, Cxy) \supset Fy)$ hold, then $(x)Fx$ also holds, we would have been faced with an infinite task, since although each block model is finite, there is no limit to the number of such models.

6. The Power of Block Models

The power of block models, and the need for the stipulation in the truth conditions for the universal quantifier that the domain D be a subset of D′, may be illustrated by considering one more formula, $(x)(\exists y)y > x$. We define $y > x$ as $(\exists z)(z \neq 0 \,\&\, Sxzy)$. To show that $(x)(\exists y)y > x$ is valid, assume for *reductio* that there is a model M such that $v_M((x)(\exists y)y > x) = F$. That is, $v_M((\exists x)(y) \sim (y > x)) = T$.

That is, there is an M′ which differs from M at most in assignment to x, such that $v_{M'}(y) \sim (y > x) = T$. That is, for all models M″ = <D″, v''>, where D′ is a subset of D″ and M″ differs from M′ at most in assignment to y, $v_{M''}(\sim(y > x)) = T$. Consequently, for all such M″, $v_{M''}(y > x) = F$.

But this is false. There will be an M″ = <D″, v''), which in its assignment function v'' differs from M′ at most in assignment to y, such that $v_{M''}(y > x) = T$. M″ simply assigns to y a linear array of blocks that is one block longer than the array that M′ assigns to x. This is possible because of the requirement that D′ is a subset of D″.

7. How Robinson Came Close to Proving the Consistency of Arithmetic in 1964

No doubt with Gödel's second incompleteness theorem in mind, Robinson remarks, on p. 234 of his (1965a), (p. 511 of the (1979) reprint), that "it is indeed a regrettable fact that no version of classical Mathematics is provably consistent". Further on in the same paper he says, "the gap due to the absence of consistency proofs for the major mathematical theories appears to be inevitable and we have learned to live with it." (p. 513 of the reprint). But in the appendix of his paper, entitled "A notion of potential truth", he introduces a concept which, if pursued far enough, would have led him to a semantic consistency proof very similar to the one put forward in the present paper. Robinson discusses relational structures ordered by inclusion, so that if A and B are any such structures, there is a structure C such that A and B are both included in C. The set of such structures is an upward directed set. He is not specific about what relations go to make up such structures, but as examples he cites (reprint, p. 523) the relations S*xyz*, P*xyz* and E*xy* of addition, multiplication and equality. Robinson explicitly introduces these relations in his book on Model Theory (1965b) pp. 26–27, published at roughly the same time. The basis of the ordering of Robinson's structures appears to be that a structure P is included in a structure Q if all the individuals of P are contained in Q, but Q may contain one or more new individuals not contained in P. He offers as an example the set N of natural numbers with the relations of addition, multiplication and equality, where the relational structures A, B, C,. are restricted to finite initial segments $N_k = \{0, 1, \ldots k\}$, $k = 0, 1, 2, \ldots$, of the natural numbers.

Robinson introduces the notion of "potential truth" of a sentence X in a structure A in words very similar to those used to define the valuation function v_M in section 4 above. He defines, successively, potential truth for atomic formulae, truth-functions, and quantified formulae. He proves that if a sentence X is potentially true in a structure A, then it is potentially true in all structures that include A. Finally, he demonstrates that X is true (not "potentially true", but true in an unqualified sense) in the union M of the directed set of all relational structures if and only if it is potentially true in every relational structure A in which it is defined. (Robinson is not explicit about the conditions under which a sentence X might *fail* to be defined in a given structure A.)

Plainly, Robinson is very close to constructing a semantic interpretation of arithmetic similar to the block semantics of this paper. What is lacking is an explicit definition of what constitutes a "model". As seen above, a model consists of a non-linguistic domain together with an assignment function, and Robinson passes over the latter. If he had recognized "assignments" as linking symbols with objects, he could have identified his "potential truth" with "truth in a model", and then gone on to define "ordinary truth" as "truth in all models". That would have opened the door to characterizing the set of mathematical truths such as S*xyz* ⊃ S*yxz* as simply "true" as opposed to "potentially true", and from there it is only a short step to proving mathematical consistency.

8. A Complete Set of Robinson-Style Axioms for Peano Arithmetic

This section contains a complete set of Robinson-style axioms for Peano Arithmetic. From them, using suitable definitions, the usual axioms containing the functional terms S*x*, *x* + *y* and *x*.*y* will be derived. The axioms R1–R10 for the system RPA, plus the Rule of Induction, are as follows:

R1. S*x*0*x*
R2. ~S*x*10
R3. (S*xyz* & S*wyz*) ⊃ *x* = *w*
R4. P*x*00
R5. P*x*1*x*
R6. (*x*)(*y*)(*z*)(∃*s*)(∃*t*)(∃*u*)(S*yzs* & S*xsu* & S*xyt* & S*tzu*)
R7. (*x*)(*y*)(*z*)(∃*r*)(∃*s*)(∃*t*)(∃*u*)(S*yzr* & P*xrs* & P*xyt* & P*xzu* & S*tus*)
R8. (∃*y*)[C*xy* & (*z*)(C*xz* ⊃ *z* = *y*)]
R9. (∃*z*)[S*xyz* & (*w*)(S*xyw* ⊃ *w* = *z*)]
R10. (∃*z*)[P*xyz* & (*w*)(P*xyw* ⊃ *w* = *z*)]
} Existence and uniqueness axioms for "successor", "sum" and "product".
Induction Rule. From |- F(0) and |- (*x*)(*y*)[(F*x* & C*xy*) ⊃ F*y*] infer |- (*x*)F*x*.

From this basis, together with the definitions given below, the axioms A1–A8 for the system PA containing the functional terms S*x*, *x* + *y* and *x*.*y* are derivable. The additional definitions required are:

Df. 1 *x* + *y* = *z* is defined as S*xyz*
Df. 2 *x* × *y* = *z* is defined as P*xyz*
Df. 3 S*x* is defined as *x* + 1.

It should be stressed that the purpose of introducing the singular terms "*x* + *y*", "*x* × *y*" and "S*x*" in definitions 1–3 is solely to make possible the derivation of the theorems of PA in the system RPA. In the system lacking these definitions, the bivalent semantics for RPA hold, and they continue to hold when the definitions are added. Using the definitions, one simply eliminates the terms *x* + *y*, *x* × *y*, and S*x*.

The derivation of the axioms A1–A8 follows.

A1	1. (*x* = *y*) ⊃ (*x* = *z* ⊃ *y* = *z*)	A pure identity thesis
A5	2. *x* + 0 = *x*	R1, Df 1
	3. ~S*x*10	R2
	4. ~(*x* + 1 = 0)	3, Df 1

A2	5. $\sim(Sx = 0)$	4, Df 3
	6. $(x = y) \supset (x + 1 = y + 1)$	Identity thesis
A3	7. $(x = y) \supset (Sx = Sy)$	6, Df 3
	8. $(Sxyz \ \& \ Swyz) \supset x = w$	R3
	9. $(Sx1z \ \& \ Sw1z) \supset x = w$	8
	10. $[(x + 1 = z) \ \& \ (w + 1 = z)] \supset x = w$	9, Df 1
	11. $[(Sx = z) \ \& \ (Sw = z)] \supset x = w$	10, Df 3
	12. $(\exists z)[(Sx = z) \ \& \ (Sw = z)] \supset x = w$	11, Pred. Logic
	13. $(Sx = Sw) \supset (\exists z)[(Sx = z) \ \& \ (Sw = z)]$	Pred. Logic with Id.
A4	14. $(Sx = Sw) \supset x = w$	12, 13
	15. $(x)(y)(z)(\exists s)(\exists t)(\exists u)(Syzs \ \& \ Sxsu \ \& \ Sxyt \ \& \ Stzu)$	R6
	16. $(\exists s)(\exists t)(\exists u)(y + z = s \ \& \ x + s = u \ \& \ x + y = t \ \& \ t + z = u)$	15, Df 1
	17. $(\exists u)[(x + (y + z) = u) \ \& \ ((x + y) + z) = u]$,	Pred. Logic with Id.
	18. $x + (y + z) = (x + y) + z$	17
	19. $x + (y + 1) = (x + y) + 1$	18
A6	20. $x + Sy = S(x + y)$	19, Df 3
	21. $Px00$	R4
A7	22. $x \times 0 = 0$	21, Df 2
	23. $(x)(y)(z)(\exists r)(\exists s)(\exists t)(\exists u)(Syzr \ \& \ Pxrs \ \& \ Pxyt \ \& \ Pxzu \ \& \ Stus)$	R7
	24. $(\exists r)(\exists s)(\exists t)(\exists u)(y + z = r \ \& \ (x \times r) = s \ \& \ (x \times y) = t \ \& \ (x \times z) = u \ \& \ t + u = s)$	23
	25. $(\exists s)[(x \times (y + z) = s) \ \& \ ((x \times y) + (x \times z) = s)]$	24
	26. $x \times (y + z) = (x \times y) + (x \times z)$	25, Pred. Logic
	27. $Px1x$	R5
	28. $x \times 1 = x$	27, Df 2
	29. $x \times (y + 1) = (x \times y) + (x \times 1)$	26
A8	30. $x \times Sy = (x \times y) + x$	29, 28 Df 3
	31. $(x + 1 = y) \equiv Sx1y$	p. Logic, Df 1
	32. $(Sx = y) \equiv Sx1y$	Df 3
	33. $(Sx = y) \equiv Cxy$	32, Df "Cxy"
	34. $(x)[Fx \supset F(Sx)] \supset (x)(y)[(Fx \ \& \ Sx = y) \supset Fy]$	Pred. Logic with Id.
	35. $(x)[Fx \supset F(Sx)] \supset (x)(y)[(Fx \ \& \ Cxy) \supset Fy]$	33, 34

Using 35, we can derive PA's induction rule from RPA's induction rule.

PA's induction rule is:

From $\vdash F(0)$ and $\vdash (x)(Fx \supset F(Sx))$ infer $\vdash (x)Fx$.

9. A Note on Finite and Infinite Models

It has been suggested by a colleague that the consistency proof using block models does not go through unless the existence of an infinite set of blocks is assumed, and that the universe could contain such an infinite set might be doubted. However, the consistency proof does not require an infinite set of blocks, as may be seen as follows.

The truth-conditions for existentially quantified formulae as given above, in section 4.1.2, are:

(iv) $v_M((\exists x)A) = T$ in model $M = \langle D, v \rangle$ iff there is a model $M' = \langle D', v' \rangle$, where D is a subset of D′ and where v' differs from v at most in assignment to x, such that $v_{M'}(A) = T$.

Since D is a subset of D′ and consequently M′ can be a larger model than M, the question arises whether it is not merely some finite structure M, but rather the infinite structure which is the collection of all block models—the universe of block models, so to speak—that is shown to satisfy RPA. Since block models are based on domains of physical cube-shaped objects, the infinite structure consisting of the set of all block models would not exist unless the universe contained infinitely many distinguishable physical items. But whether the universe contains infinitely many things, or only finitely many things, is an empirical matter. Without an infinity of physical objects, does the absolute consistency proof go through?

There are two replies to this. First and foremost, at no point in the proof that every axiom of RPA is semantically valid, i.e. true in every non-empty block model, is an infinite model presupposed. All block models have finite domains, there being no largest one. There exists no limit to the size of block domains, without any single domain being infinite. To address the question of whether the physical universe could contain a *potential* infinity of blocks, imagine the following thought experiment.

Take a single cube of matter, measuring 1 × 1 × 1. (The word "matter" is used in a 17th century sense, before the advent of atomic theory.) The units of length are irrelevant: they could be meters, or kilometers, or whatever. Imagine that the cube is bisected along each of its three dimensions, producing 8 smaller cubes, each ½ × ½ × ½. Continue this process, through 64, 512, 4096, ... cubes. Out of a single 1 × 1 × 1 cube, a set of smaller and smaller cubes can be created, and the size of this set has no numerical upper limit. Each step produces a larger, but finite, number of blocks. A set of blocks that is actually infinite is never produced, nor is it necessary for the consistency proof. Block models differ sharply from the "standard model" of arithmetic, based on an infinite set of integers. The latter does not yield a non-circular arithmetic consistency proof, as is argued in section 5 above, axiom R6.

It might be asked, is it not an easy step from the predicate "...is a block model" to the *set of all block models*, which is infinite? Yes, by using the (unrestricted) axiom schema of comprehension, which states that to every property there corresponds a set. But this leads to Russell's paradox, and Frege's lament that arithmetic totters. The consistency proof neither needs nor recognizes the set of all block models, nor does it require a domain consisting of an actual infinity of physical blocks.

References

Church, Alonzo (1956), *Introduction to Mathematical Logic*, vol I, Princeton: Princeton University Press.

Goodstein, R.L. (1961), *Mathematical Logic*, Leicester: Leicester University Press.

Lewis, David (1986), *On the Plurality of Worlds*, Oxford and New York: Blackwell.

Mendelson, Elliott (1964), *Introduction to Mathematical Logic*, Princeton: Van Nostrand.

Robinson, Abraham (1965a), "Formalism 64", in *Proceedings of the International Congress for Logic, Methodology and Philosophy of Science*, Jerusalem 1964, North Holland Pub. Co., Amsterdam, pp. 228-246. Reprinted in *Selected Papers of Abraham Robinson*, vol. 2, ed. Keisler et al., Yale University Press 1979, pp. 505–23.

Robinson, Abraham (1965b), *Introduction to Model Theory*, Amsterdam: North-Holland.

van Fraassen, Bas C. (1980), *The Scientific Image*, Oxford and New York: Oxford University Press.

{ 2 }

Can a Turing Machine Know That the Gödel Sentence Is True?

A Turing machine can *know* what it can *prove*, that is, deduce from axioms using well-defined rules of inference. Its axiomatic database may be large, and its rules of inference may be efficient and powerful. But for a Turing machine, *knowability = provability.* When truth and provability part company, as in the case of Gödel's incompleteness theorem, a machine can cope with one but not with the other.

The short answer to the question in the title is "no", because if *Peano arithmetic* (PA) should turn out to be inconsistent, the Gödel sentence *G* would be false not true. The truth value of *G* depends on an unknown: the consistency of PA. In any case, truth and provability diverge: for if PA is consistent, *G* is true and improvable; and if inconsistent, provable and false. On the other hand, conditionalizing on PA's consistency yields "if . . . then" sentences that are unconditionally true or false, and here the connection between truth and provability is more subtle. In section 2.2, two conditional sentences are examined:

A1: If PA is consistent, then *G* is not provable.
A2: If PA is consistent, then ~*G* is not provable.

Both sentences are true, the difference being that the formal version of Al in the language of PA is a theorem, whereas the formal version of A2 to the best of our knowledge is not. Although a "PA Turing machine", programed to run through and check proofs in PA, will be able to enumerate all PA theorems, and hence will *know* A1 in the sense that it can prove it, the machine, unlike a human being, will be incapable of recognizing, grasping, or otherwise perceiving the truth of A2. The concept of "unprovable but true" is beyond it. Or so I shall argue.

The author wishes to thank Anil Gupta, Allen Hazen, Zlatan Damnjanovic, and Alasdair Urquhart for helpful comments.

There have been two recent attempts, by John Lucas and Roger Penrose,[1] to use Gödel's theorem to show that any computational device is limited, compared to human beings, in what it can know or understand. Their arguments, which if correct would imply that the program of strong AI was ultimately not viable, have provoked vigorous rebuttals from other philosophers.[2] I shall argue along different lines. I focus on the sharp distinction between truth and provability and show that whether PA is consistent or inconsistent, the set of provable sentences differs from the true sentences. I then take up an example of a sentence in PA which is true, but which cannot be categorized as true by a PA Turing machine. The same holds for Turing machines with richer axiomatic bases; for each of them there will be at least one proposition which a human can know to be true, but which the machine lacks the capacity to recognize. At every level, there will be specific cases in which human mental powers outstrip the capabilities of Turing machines.

1. Gödel's Theorem

Among the things that Gödel showed is that, if PA is consistent, the concept of proof is representable in PA. This means that, if *Pr* is the set of ordered pairs $\langle x,y \rangle$, where x is the Gödel number (GN) of the proof of the well-formed formula with GN y, then there is an open PA formula:

$$\underline{\quad\quad} x \underline{\quad\quad} y \underline{\quad\quad}$$

which is such that for all x and y, $\langle x,y \rangle \in Pr$ if and only if:

$$\vdash \underline{\quad\quad} \bar{x} \underline{\quad\quad} \bar{y} \underline{\quad\quad}$$

where $\bar{x}$ is x in Peano notation, that is, $0''''\ldots'$.

This is the definition of *weak* representability. So-called *strong* representability[3] means that, for any x and y, if $\langle x,y \rangle \in Pr$ then $\vdash \underline{\quad\quad} \bar{x} \underline{\quad\quad} \bar{y} \underline{\quad\quad}$, and if $\langle x,y \rangle \notin Pr$ then $\vdash \sim(- \underline{\quad\quad} \bar{x} \underline{\quad\quad} \bar{y} \underline{\quad\quad})$. Recursively enumerable sets are weakly representable; decidable sets are strongly representable. Since

---

[1] Lucas, "Minds, Machines and Gödel," *Philosophy* 36 (1961): 112–27; Penrose, *The Emperor's New Mind* (New York: Oxford University Press, 1989), *Shadows of the Mind* (New York: Oxford University Press, 1994).

[2] See Paul Benacerraf, "God, the Devil, and Gödel," *The Monist* 51 (1967): 9–32; David Lewis, "Lucas against Mechanism," *Philosophy* 44 (1969): 231–33, and "Lucas against Mechanism II," *Canadian Journal of Philosophy* 9 (1979): 373–76; G. L. Bowie, "Lucas' Number Is Finally Up," *Journal of Philosophical Logic* 11 (1982): 279–85; Hilary Putnam, review of *Shadows of the Mind*, by Roger Penrose, *The New York Times Book Review*, November 20, 1994, 7; also numerous commentators in *Behavioral and Brain Sciences* 13 (1990): 655–92.

[3] See Moshe Machover, *Set Theory, Logic and Their Limitations* (New York: Cambridge University Press, 1996), 223.

it is a routine clerical matter to determine which pairs <*x,y*> belong to *Pr*, *Pr* is a decidable set and is strongly represented by ____*x*____*y*____. On the other hand, although the set *Th* of Gödel numbers of theorems of PA is enumerable because proofs are enumerable, *Th* is not decidable because the nontheorems of PA are not enumerable. The open formula that weakly represents *Th* is (∃*x*)(____*x*____*y*____).

Gödel also noted that the set Sub of triples <*w,y,z*> is decidable, where *w* is the GN of the formula which results from substituting the number *y* (in Peano notation) for all instances of free variables in the formula with GN *z*. Furthermore, *w* is a function of *y* and *z*, so that the open formula that represents Sub can be written in the form $w = \text{sub}(y,z)$. Construct the formula:

$$\text{A3}: \sim(\exists x)(\_\_\_\_x\_\_\_\_\text{sub}(y,z)\_\_\_\_)$$

and let the GN of A3 be *g*. Then the Gödel formula *G* is:

$$G: \sim(\exists x)(\_\_\_\_x\_\_\_\_\text{sub}(\bar{g},\bar{g})\_\_\_\_)$$

Since sub(*g,g*) is the GN of *G*, *G* says of itself that it is not a theorem of PA.

Is *G* provable in PA? Suppose, for reductio, that it were, and let *h* be the GN of its proof. Then $< h,\text{sub}(g,g) > \in Pr$. By representability:

$$\text{A4}: \vdash(\_\_\_\_\bar{h}\_\_\_\_\text{sub}(\bar{g},\bar{g})\_\_\_\_)$$

But from $\vdash G$ we derive:

$$\text{A5}: \vdash (x)\sim(\_\_\_\_x\_\_\_\_\text{sub}(\bar{g},\bar{g})\_\_\_\_)$$

whence by universal instantiation:

$$\text{A6}: \vdash \sim(\_\_\_\_\bar{h}\_\_\_\_\text{sub}(\bar{g},\bar{g})\_\_\_\_)$$

On the assumption that PA is consistent, A4 and A6 cannot both hold, and consequently *G* cannot be a theorem.

2. Truth and Provability in Peano Arithmetic

If *G* says that *G* is not provable in PA, and if *G* is in fact not a PA theorem, then it would appear that what *G* says is true. *G* would be true-but-unprovable. But this does not by itself show that human beings can know something—the truth of *G*—that Turing machines cannot know, since both the truth and the unprovability of *G* are predicated upon the assumption that PA is consistent. If PA should turn out to be inconsistent, *G* would be not true-and-unprovable, but provable-and-false.

There are therefore two alternatives: if PA is consistent, *G* is un-provable and true; if PA is inconsistent, *G* is provable and false. Whichever alternative holds, provability and truth part company. Either some truths are not theorems, or some theorems are not true. The domain of expertise of a Turing machine lies in the area of proof and provability, not in the area of truth. Human beings, on the other hand, are acquainted with both proof and truth, and also know of cases where the two diverge.

It might be argued, despite the above, that, if we conditionalize on PA's consistency, truth and provability come together again. Thus, on the assumption that PA is consistent, *G* is true and unprovable, but at the same time the conditional sentence A1 is true and, as we shall see, provable in PA. Similarly:

A7: If PA is inconsistent, then *G* is provable.

is also true and provable. If, it is claimed, we focus on conditional sentences, truth and provability coincide once more and Turing machines can know (=prove) whatever human beings can know (=prove).

The details of this claim must be examined. Denote the Gödel number (in Peano notation) of a well-formed formula *A* by [*A*], define the expression Bew[*A*], which says that *A* is provable, as $(\exists x)(\underline{\quad} x \underline{\quad} [A] \underline{\quad})$, and note that *G* is equivalent to ~Bew[*G*]. The assertion that PA is consistent may be symbolized in PA by a number of equivalent expressions, such as ~Bew[0 = 0′]. I abbreviate the latter by "Cons(PA)." Then A1 is written in the language of PA as:

$$\text{A8}: \text{Cons}(\text{PA}) \rightarrow \sim \text{Bew}[G]$$

or equivalently as:

$$\text{A9}: \text{Cons}(\text{PA}) \rightarrow G$$

Proving A9 as a theorem of PA is tantamount to proving Gödel's second theorem, which states that Cons(PA) is not provable in PA unless PA is inconsistent. (If Cons(PA) *were* a PA theorem, then, by A9, *G* would also be a theorem, which is known not to be the case.) In his original 1931 paper, Gödel does not prove this second theorem; a proof is to be found in David Hilbert and Paul Bernays.[4] Assuming that this proof can be fully formalized in PA, A9 is both true and provable in PA.

It follows from the above that in the case of the conditional sentences A8 and A9, truth and provability coincide, and a Turing machine can *know* the

---

[4] *Grundlagen der Mathematik*, vol. 2 (Berlin: Springer, 1939). See Stephen Cole Kleene, *Introduction to Metamathematics* (Amsterdam: North-Holland, 1952), 211; and Elliott Mendelson, *Introduction to Mathematical Logic* (Princeton: Van Nostrand, 1964), 148.

truth of A9 because it can prove A9 as a theorem. But a different result holds for the formal version of A2. This is:

$$\text{A10}: \text{Cons}(\text{PA}) \rightarrow \sim \text{Bew}[\sim G]$$

Moving from A9 to A10 yields a candidate for the true-but-unprovable sentence that we seek.

Is A10 true? Suppose, for reductio, that PA is consistent and that ~*G* is a theorem. Then $[\sim G] \in Th$, and, from the consistency of PA we infer that $[G] \notin Th$. At the same time, since ~*G* is equivalent to $(\exists x)(\_\_\_x\_\_\_[G]\_\_\_)$, from $\vdash \sim G$ we derive $(\exists x)(\_\_\_x\_\_\_[G]\_\_\_)$. Because the open formula $(\exists x)(\_\_\_x\_\_\_y\_\_\_)$ weakly represents *Th*, it follows that $[G] \in Th$, which completes the reductio. Hence A10 is true.[5]

The informal argument just presented cannot be converted into a formal proof of A10's theoremhood, and there are good reasons to believe that A10 is in fact unprovable in PA. Because of the equivalence of ~*G* to Bew[*G*], A10 can be written as:

$$\text{A11}: \text{Cons}(\text{PA}) \rightarrow \sim \text{Bew}[\text{Bew}[G]]$$

In the logic of provability, one of the laws governing provability as a modal notion is:

$$\text{A12}: \text{Bew}[X] \rightarrow \text{Bew}[\text{Bew}[X]]$$

("If *X* is provable, then it is provably provable").[6] Instances of A12 are theorems of PA. Since the converse of A12 does not hold, the consequent ~Bew[Bew[*G*]] of A11 is a net strengthening of the consequent ~Bew[*G*] of A8, and it is unlikely that the long and complex proof of A8 in PA could be reworked so as to yield a proof of A11. Not impossible, perhaps, in the absence of a knock-down independence proof, but highly unlikely.

The upshot is that a Turing machine programed to run through proofs in PA will almost certainly never classify A10 as a theorem (unless, of course, PA is inconsistent), but neither will it be capable of assigning it to the special category of "unproven but true." Recognition of the existence of such a category therefore marks a difference in principle between human and machine thinking.

Despite the above reasoning, there will still be those who wish to maintain that somehow, in some fashion, a Turing machine can recognize the truth of

---

[5] It is frequently stated that proving the nontheoremhood of ~*G* requires the assumption that PA is not only consistent but *w*-consistent (see, for example, Mendelson, 143). The argument given in the text shows, however, that the weaker assumption of ordinary consistency suffices.

[6] See George Boolos, *The Logic of Provability* (New York: Cambridge University Press, 1993), 15.

A10. There are three possible strategies for arriving at this conclusion. The first is to show that "truth," like "provability," is in some sense expressible by a formula of PA. The second is to formalize the notion of truth by introducing a new primitive sentential operator *T*, the truth operator, and show that in PA, suitably enriched with extra axioms, *T*(A10) and similar theorems may be derived. The third is to hold that adding new axioms to PA, without the operator *T*, allows A10 to be proved, and that in stronger and stronger extensions of PA, a Turing machine can derive every proposition a human can know to be true. I shall argue that all these strategies fail.

3. Expressing Truth by an Arithmetical Formula

Tarski's theorem states that the set *Tr* of Gödel numbers of well-formed formulas of PA which are true in the standard interpretation is not "arithmetical", that is, there is no open sentence ____*x*____ of PA such that, for any *x*, *x* belongs to *Tr* if and only if the closed sentence ____$\bar{x}$____ is true in the standard interpretation.[7] The notion of "truth" in PA is consequently on quite a different footing from that of "provability". The latter concept is represented by an open arithmetical formula; no analogous formula expresses (much less represents) the former, thus reinforcing the hypothesis that "truth", though meaningful to humans is a closed book to a Turing machine.

4. Introducing the *T* Ooperator

The second possible way of supplementing a Turing machine's ability to deal with proof by an ability to deal with truth is to introduce an explicit truth operator *T.* With suitable *T* axioms, the machine might then be able to prove propositions like:

$$\text{A13}: T(\text{If PA is consistent, then} \sim G \text{ is not provable})$$

For steps in the implementation of this strategy, see, for example, Solomon Feferman.[8] There are, however, two worries. The first is that the *T* axioms may be able to do no better than supply, for any already-proved theorem *X*, a new theorem *T(X)*. If *X* were itself not a theorem of PA, as is probably the case with A10, then *T(X)* would also fail to be a theorem. If this were so, introduction of

---

[7] Alfred Tarski, "The Concept of Truth in Formalized Languages," in *Logic, Semantics, Metamathematics* (New York: Oxford Clarendon Press, 1956), 152–278. See Mendelson, p. 151; Machover, p. 236.

[8] "Reflecting on Incompleteness," *Journal of Symbolic Logic* 56 (1991): 1–49, here pp. 13 ff.

the T operator would do nothing to correct a Turing machine's inability to cope with unprovable but true propositions.

The second worry is more general. T axioms may increase a Turing machine's deductive powers in the sense of giving it more theorems it can prove, that is, broadening the extension of "proof." But how is this relevant to "truth"? Take the analogy of non-Euclidean geometries. If a Turing machine is programed to accept the Gödel numbers of formal axiomatic proofs in Euclidean geometry, and if the parallel postulate is then replaced by Lobatchevsky's postulate and the machine is reprogramed, it will generate a different set of proofs and a different set of theorems. But will the reprogramed machine categorize the old parallel postulate as "true but unprovable"? Or as "formerly provable but false"? This seems unlikely. Between provability and truth there yawns an unbridgeable gulf.

5. Stronger and Stronger Extensions of PA

The third way of getting a Turing machine to accept A10 as true, without introducing the operator T, is to strengthen PA and deduce A10 from new axioms.

But again, there is a difficulty. No matter what system L of axioms and rules the machine is programed to use, and provided that L is at least as strong as PA so that "proof" and "substitution" are representable, there will always be a sentence analogous to A10 which is true, but for which the Turing machine will be unable to find a proof. In Zermelo-Frankel set theory, for example, an appropriately programed machine will be able to construct a Gödel sentence G_{ZF} and a ZF-analogue of A10:

$$\text{A14}: \text{Cons}(\text{ZF}) \rightarrow \sim \text{Bew}[\sim G_{ZF}]$$

The machine will be able neither to prove this sentence, nor to recognize it as true. A human being, on the other hand, can know that every such sentence, in every extension of PA, is true.

To sum up, what is the source of a Turing machine's inability to recognize the truth of A10? Can the inability be done away with by moving to more powerful programs? Let us start with PA.

A Turing machine, programed to run through and list all proofs valid according to the axioms and rules of PA, cannot recognize A10 to be true if it cannot prove it. For such a machine, there are only two categories of well-formed formulas: those which are theorems and those for which no proof has yet been found. (If PA is discovered to be inconsistent, the second category will turn out to be empty.) For the machine, the category "not yet proven but true" does not exist.

What now of stronger machines, equipped with more powerful axiomatic bases and able to generate larger sets of theorems? What of machines equipped with the truth operator *T*? What of machines with huge data banks? It is difficult to see why the situation, for them, is any different from what it is for a PA machine. There still remain the two categories of theorems and nontheorems, and there still remains, for each of them, a sentence analogous to A10. This sentence is unproven but true. The Achilles heel of all Turing machines is that for none of them do the categories of "true but not a theorem" and "theorem but not true" exist. The gulf that separates human from machine thinking is the gulf that separates provability from truth.

References

Benacerraf, Paul. "God, the Devil, and Gödel." *The Monist 51* (1967): 9–32.

Boolos, George. *The Logic of Provability.* New York: Cambridge University Press, 1993.

Bowie, G. L. "Lucas' Number Is Finally Up." *Journal of Philosophical Logic 11* (1982): 279–85.

Feferman, Solomon. "Reflecting on Incompleteness." *Journal of Symbolic Logic 56* (1991): 1–49.

Hilbert, David, and Paul Bernays. *Grundlagen der Mathematik*, vol. 2. Berlin: Springer, 1939.

Kleene, Stephen Cole. *Introduction to Metamathematics.* Amsterdam: North-Holland, 1952.

Lewis, David. "Lucas against Mechanism." *Philosophy* 44 (1969): 231–33.

——. "Lucas against Mechanism II." *Canadian Journal of Philosophy* 9 (1979): 373–76.

Lucas, J. R. "Minds, Machines and Gödel." *Philosophy* 36 (1961): 112–27.

Machover, Moshe. *Set Theory, Logic and Their Limitations.* New York: Cambridge University Press, 1996.

Mendelson, Elliott. *Introduction to Mathematical Logic.* Princeton: Van Nostrand, 1964.

Penrose, Roger. *The Emperor's New Mind.* New York: Oxford University Press, 1989.

——. *Shadows of the Mind.* New York: Oxford University Press, 1994.

Putnam, Hilary. Review of *Shadows of the Mind*, by Roger Penrose. *The New York Times Book Review*, November 20, 1994, 7.

Tarski, Alfred. *Logic, Semantics, Metamathematics.* New York: Oxford Clarendon Press, 1956.

{3}

On "Seeing" the Truth of the Gödel Sentence

What human beings can know to be true, and what Turing machines can prove, are two different things. In every formal system at least as strong as Peano arithmetic (PA) there is a formulable Gödel sentence which is unprovable, and which a Turing machine programmed to run through possible proofs and generate the theorems of the system will never produce as part of its output, unless the formal system in question is inconsistent. Human beings, on the other hand (again under the assumption that the system is consistent), can recognize the Gödel sentence to be true. The machine can assign the sentence to the category "unproven", but for it the category of "unproven but true" does not exist.

In a short paper with the same title as this one, George Boolos comes to essentially these conclusions, but with an important difference (Boolos 1990). While acknowledging that if PA is assumed to be consistent, the Gödel sentence G for PA is both true and unprovable-in-PA, Boolos argues that the same does not hold for stronger systems like Zermelo-Fränkel set theory (ZF). Though there is a sense in which we can "see" the truth of G in PA, Boolos offers reasons for thinking we cannot do the same for the Gödel sentence G′ of ZF. I shall argue, on the other hand, that PA and ZF are perfectly analogous in this respect, and that our ability to recognize the truth of the Gödel sentences for PA and all stronger systems reveals a sharp dividing line between human and machine thinking. This latter conclusion is derived and discussed in McCall (1999).

The construction of Gödel sentences for PA and other systems is not difficult, given a system of Gödel numbering which enables these systems to talk about their own syntax. Let Pr be the set of ordered pairs of numbers $<x,y>$, where x is the Gödel number (GN) of a proof of the wff with GN y. It is a routine matter for a Turing machine to go systematically through all pairs of numbers and determine which are, and which are not, members of Pr. Pr is a *decidable* set, and Gödel showed that there is an open formula ----x----y---- of PA containing two free variables which *represents* the set Pr in the following sense: for all *x,y* the pair <*x,y*> belongs to Pr iff, in PA, |- ---- $\overline{x}$ ---- $\overline{y}$ ---- (where the numeral $\overline{x}$ is the number *x* in Peano notation, i.e. 0 followed by *x* primes).

A second decidable set which Gödel also showed to be representable in PA is the set Sub of ordered triples <*w,y,z*>, where *w* is the GN of the formula which results from substituting the number *y* (in Peano notation) for all free variables in the formula with GN *z*. Given *y* and *z*, the number *w* is uniquely determined, hence *w* is a function of *y* and *z* and the open formula which represents Sub in PA can be written in the form *w* = sub(*y,z*).

Write down the formula

$$(1)\ \sim(\exists x){-}{-}{-}{-}x{-}{-}{-}{-}\mathrm{sub}(y,z){-}{-}{-}{-}.$$

What (1) says is that there is no number *x* which is the GN of a proof of the wff with GN sub(*y,z*). Let the GN of (1) be *g*. Then the Gödel sentence G for PA is:

$$\text{G.}\ \sim(\exists x){-}{-}{-}{-}x{-}{-}{-}{-}\mathrm{sub}(\overline{g},\overline{g}){-}{-}{-}{-}.$$

An important part of "seeing" the truth of the Gödel sentence is seeing, i.e. understanding, what it is that G asserts. G says that there is no GN of a proof of the wff with GN sub(*g,g*), hence that this wff is unprovable. But the wff with GN sub(*g,g*) is G itself, hence G says of itself that it is not a theorem of PA.

Is G true? A simple argument shows that it is, if PA is assumed to be consistent. Suppose, for *reductio*, that G is a PA theorem, and let *h* be the GN of its proof. If we denote the GN of any formula A by [A], then <*h*,[G]> $\in$ Pr, and since ----*x*----*y*---- represents Pr, we have:

$$(2)\ \vdash{-}{-}{-}{-}\overline{h}{-}{-}{-}{-}{-}[\mathrm{G}]{-}{-}{-}{-}.$$

But at the same time, since G is a PA theorem, we also have:

$$(3)\ \vdash\ \sim(\exists x){-}{-}{-}{-}x{-}{-}{-}{-}\mathrm{sub}(\overline{g},\overline{g}){-}{-}{-}{-}.$$

Since [G] = sub(*g,g*) = sub($\overline{g}$, $\overline{g}$), (3) yields

$$(4)\ \vdash(x)\ \sim\ ({-}{-}{-}{-}x{-}{-}{-}{-}[\mathrm{G}]{-}{-}{-}{-}),$$

whence by universal instantiation

$$(5)\ \vdash\ \sim({-}{-}{-}{-}\overline{h}{-}{-}{-}{-}[\mathrm{G}]{-}{-}{-}{-}),$$

which contradicts (2). If PA is consistent, G cannot be a PA theorem, and what G says is true.

To "see the truth" of G, assuming that PA is consistent, one must (i) understand what assertion G makes, and (ii) be able to follow the *reductio* argument of the previous paragraph. Neither of these tasks is difficult: logic students in all parts of the world regularly see that G is both true and unprovable in PA. But a Turing machine programmed to identify pairs <*x,y*> which belong to Pr, and thereby progressively generate a list of all PA theorems, will never output G, and will therefore never classify G as true.

A *caveat* needs to be entered here, namely that the entire argument up to this point has been predicated on PA's consistency. If PA should turn out to be inconsistent G would be false not true, and those claiming to have "seen" its truth would have to recant. But in fact nothing of substance in the argument for the difference between human and machine thinking is lost, for even if PA is inconsistent, truth and provability still part company. We argue by cases:

Case 1. PA is consistent. G is unprovable, but true.
Case 2. PA is inconsistent. G is provable, but false.

Whichever alternative holds, truth and provability fail to coincide, and PA contains true non-theorems or false theorems. Human beings are familiar with the distinction between truth and theoremhood, but Turing machines cannot look beyond their own output.

Moving from PA to ZF, Boolos is pessimistic about the chances of humans ever being able to see the truth of the Gödel sentence G′ for ZF. I shall argue, by contrast, that the situation is in all relevant respects the same for ZF as it is for PA. Let Pr(ZF) be the set of pairs <*x,y*> where *x* is the GN of a ZF-proof of *y*. Since Pr(ZF) is decidable it is representable in PA: let....*x*....*y*....be the PA formula that represents it. Similarly let $x = \text{sub}'(y,z)$ be the PA formula that represents substitution in ZF. Then the Gödel sentence G′ for ZF is:

$$\text{G}'.\ \sim(\exists x)....x....\text{sub}'(g',g')...., \text{ i.e.}$$

$$\text{G}'.\ \sim(\exists x)....x....[\text{G}'].....$$

G′ says that there is no ZF-proof of G′, i.e. that G′ is not a ZF theorem. If ZF is consistent, is G′ true? Well, assume for *reductio* that ZF is consistent and that G′ is false, i.e. that G′ is a ZF theorem. These assumptions yield a contradiction using the argument of three paragraphs back. Therefore if ZF is consistent G′ is true, and can be seen to be true-and-unprovable by humans though not by ZF Turing machines.

The reason why Boolos is sceptical of our ability to see the truth of the Gödel sentence for ZF is that he focusses not on G′, but on a sentence formally equivalent to it, namely the ZF sentence Cons (ZF) which asserts that ZF is consistent. There are various ways to formulate Cons(ZF), one of which is to take it as ~(∃*x*)….*x*….[0 = 1]…., which asserts that 0 = 1 is not provable in ZF. The sentence:

$$(6)\ \text{Cons(PA)} \rightarrow \text{G}$$

is a PA theorem, and presumably also

$$(7)\ \text{Cons(ZF)} \rightarrow \text{G}'$$

is a ZF theorem, which (since the converse of (7) is trivial) makes G′ provably equivalent to Cons(ZF).

Here is Boolos's argument why we can't "see the truth" of the Gödel sentence for ZF, even though we may conceivably be able to see the truth of the Gödel sentence for PA. In Boolos's words:

> What we can see the truth of is this conditional proposition: the Gödel sentence for ZF is ZF-unprovable (and therefore true) *if* ZF is consistent. We cannot see that the Gödel sentence is true precisely because we cannot see that ZF is consistent. We may hope or believe that it is, but we do not know it, and therefore cannot see it (Boolos 1990, p. 655).

This argument is unconvincing because, in replacing G′ by its equivalent Cons(ZF) as the Gödel sentence, it leaves out the critical dependency of the truth-value of G′ on ZF-consistency. No one knows whether ZF is consistent or inconsistent, and therefore no one knows whether G′ is true or false. There is, certainly, a fact of the matter as to the consistency of ZF, and the truth-value of G′ depends on this unknown fact. If ZF is consistent, G′ is true, and if ZF is inconsistent, G′ is false. Since we understand what G′ says, namely that G′ is unprovable, we also understand why G′ has a truth-value which depends on the consistency of ZF (bearing in mind that in an inconsistent system all wffs are theorems). Whether ZF is consistent or inconsistent, we can see in each case which truth-value G′ has.

The one-way dependency of the truth-value of G′ on ZF's consistency disappears when one replaces G′ by its equivalent Cons(ZF). (One would scarcely say that the dependency was the other way round, and that the consistency of ZF depended on whether G′ was true.) In his introduction to *The Logic of Provability*, Boolos indicates that there may be difficulties in giving clear truth-conditions for self-referential sentences like G and G′, which assert that they themselves are unprovable (Boolos 1993, p. xxx). Boolos says that with the aid of the fixed point theorem we can see how to replace self-referential sentences of arithmetic (like G and G′) with equivalent sentences which involve no self-reference (like Cons(ZF)). But to do so is to miss an important facet of Godel's incompleteness theorem, one which moreover has implications for the attempt to give a philosophically adequate account of the differences between human and machine thinking. To give clear truth-conditions for the self-referential sentences G and G′ is relatively uncomplicated, once the dependence of their truth-values on the consistency of their respective parent systems is acknowledged. Given this dependence, one can "see" their truth or falsity without difficulty, in contrast to the dark unknown status of PA's and ZF's consistency.

To conclude, what are the implications of human beings' ability to recognize the truth-status of the Gödel sentences in addition to their theoremhood-status, the latter being the sole area of expertise of Turing machines? No doubt, if a true sentence U were missing from the output of a Turing machine V, or from the output of a universal Turing machine on input W, one could go to a stronger system X, instantiated by a stronger machine or program, and add U as a new axiom. But then a new Gödel sentence Z would be constructible for system X, which would in turn be true-but-unprovable in X unless X were inconsistent. In short, the same situation would be repeated.

For every formal system, and every computational device which instantiates that system, including modern computers with large databanks, there is a Gödel sentence the status of which is clear to human beings, but opaque to the computational device. Whether recognizing or "seeing" the truth-value of Gödel sentences requires intellectual powers so different from those of a Turing machine as to qualify as "non-algorithmic" thinking, as is claimed by Roger Penrose (1989, pp. 416–418) is a further question, not addressed here. I stop at this conclusion: no intelligent program can model all of human reasoning while at the same time being run on a Turing machine.

References

Boolos, George 1990: "On 'seeing' the truth of the Gödel sentence" *Behavioral and Brain Sciences* 13, pp. 655–656, reprinted in Boolos (1998).

—— 1993: *The Logic of Provability.* Cambridge and New York: Cambridge University Press.

—— 1998: *Logic, Logic, and Logic.* Cambridge, MA: Harvard University Press.

McCall, Storrs 1999: "Can a Turing machine know that the Gödel sentence is true?" *The Journal of Philosophy* 96, pp. 525–532.

Penrose, Roger 1989: *The Emperor's New Mind.* Oxford and New York: Oxford University Press.

{4}

How to Make a Decision

1. Introduction

Decision-making is a characteristically human activity, one that most of us engage in every day. Bacon and eggs for breakfast, or just toast? Tea or coffee? Shall I walk to work, take the bus, or ride the Metro? Should I start to write that paper on free will for the conference next September, or wait until I've had a chance to read the latest research on neurophysiology of acts of choice? Decisions, decisions, decisions. Sometimes it's enough to make us want to follow Paul Gauguin to the South Seas.

Nevertheless, careful study of the decision-making process can yield a store of philosophical insights, many of them new and many of them able to shed light on the perennial question of what it is to be human. Or so I believe. This chapter is an attempt to push aside another inch or two of the curtain of unknowing, the curtain which separates us from ourselves and impedes self-knowledge. As philosophers, we all wrestle with our personal daemons, and two of mine are the daemons of deliberation and decision.

2. Deliberation

The process of deliberation, Aristotle's *bouleusis*, is divided into three stages, which I shall designate as *choice-set formation, evaluation*, and *choice*. Let me illustrate with the example of Francine, a student graduating from UQ at Trois-Rivieres, who must make up her mind as to what to do next year. Francine has been accepted into environmental studies at the Sorbonne (A), and into a PhD program in social anthropology at Chicago (B), and finally has the chance to do volunteer work in rural development in Africa (C). These options, together with the default option D ("None of the above"), constitute Francine's choice set {A, B, C, D}.

Once her choice set has been formed, Francine deliberates. This phase of the process can last for hours, or days, or weeks, and consists of weighing the pros and cons of each of the various options against one another. Paris for example is a city that Francine has always wanted to live in, but at the same time it is expensive, and the small bourse available to her will not be sufficient. Chicago on the other hand has made a generous offer of postgraduate support, but the distinguished scholar whom Francine wanted to work with has just moved from Chicago to California. Francine's father is a development officer with the World Bank who specializes in agriculture, but Francine feels nervous about spending a year in Africa without her partner Ghislain, who has a pathological fear of snakes. I give the name "deliberation reason", or "delib-reason" for short, to each of the reasons which a particular deliberator has for choosing or not choosing one of the options A, B, C,.... Thus in Francine's case love of Paris is a positive delib-reason in relation to alternative A, while Ghislain's fear of snakes is a negative delib-reason in relation to C. In the middle phase of the deliberative process, which I call "evaluation", the deliberator assesses the relative weights of the different sets of positive and negative delib-reasons, and weighs one set against the others.

The comparative weighing process involved in deliberative evaluation is one we perform so frequently, e.g. in shopping at the supermarket where nutritional value, taste and price are balanced against one another, that we forget how complicated it really is. Before delib-reasons can be *weighed* they must be (subjectively) *weighted*. Thus before the reasons for purchasing delicious but expensive leeks can be weighed against the reasons for cheap but ordinary carrots, a weight of some kind has to be placed on delectability, and a weight of some kind has to be assigned to price. A shopper with resources of only \$20 will assign different weights to price and delectability than a shopper with \$200. But though *weighting* of delib-reasons comes first, weighting gets us nowhere without comparative *weighing*, in which the exotic taste and high price of leeks are weighed against the run-of-the-mill taste and low price of carrots. Once the process of weighting and weighing is complete, the process of evaluation comes to an end with the creation of an *ordered list of preferences*. Let us suppose for example that Francine, after a week of hard thought, comes up with the evaluative list CAB: Africa, Paris and Chicago in that order.

For many philosophers, especially rational choice theorists and decision theory specialists, who focus on problems like Arrow's paradox, the prisoner's dilemma and Newcomb's problem, the production of a single agreed-on ordered list would be the end of the matter. If after due deliberation, possibly in consultation with her parents, professors and Ghislain, Francine concludes that her preferences are Africa, Paris and Chicago in that order, then one would think that the process of decision-making ought to be at an end. But softly. In the model I am proposing there is one more step in the process after evaluation, and that is *choice*, or what Aristotle calls *prohairesis*. We may imagine that Francine has arranged three letters on the table in front of her, one accepting the Sorbonne offer, one the Chicago offer, and one the Africa offer. All she has to do is sign one of them, put it in an envelope, and mail it. Since Francine now has her ordered list CAB, it would appear as if the process of deliberation was over. Certainly all decision theorists would agree that it was.

Plato, in the *Meno* 77B–78A, adopts a very similar position when Socrates asks Meno whether it is possible to desire what is dishonourable, i.e. something bad or evil. Meno believes that this is indeed possible. But, asks Socrates, does someone who apparently desires evil do so knowing it to be evil, or believing it to be good? Socrates quickly convinces Meno that people always act in accordance with what they *think* is good, even though they may be mistaken in that belief. Consequently all evil-doing is based on ignorance of the good, and no one who knows the good will ever voluntarily do evil. The relevance of this to rational choice theory is that no one who constructed a list of preferred options, based on his or her honest assessment of what was best, would ever choose an alternative other than the one at the top of the list. In this sense rational choice theorists follow Plato.

However, as we know, another equally famous philosopher notoriously differed from Plato on this point. For Aristotle it is perfectly possible to know the good but not do it, and although those who act in this lamentable and un-Platonic way may not be fully rational, they are still human. Which philosopher holds up the mirror in which we see ourselves? Is it Plato or Aristotle? Suffice it to say that Aristotle has shown us that there can be a gap between knowing what is best to do, and doing it. For Plato and the decision theorists on the other hand there is no gap, and intelligent agents will always follow their established preferences. To do otherwise would be irrational. But are human beings essentially rational? Aristotle, the realist, could perhaps see something here that Plato, the idealist, could not. In the account of decision-making put forward in this chapter, identifying one's preferred course of action, and deciding to do it, are two distinct things.

In saying that evaluation and choice are distinct I am not saying that deliberators frequently jettison their preference lists and choose irrationally, under the sway of *akrasia*. On the contrary, in the vast majority of cases we act rationally, implementing the option we judge best. But what Aristotle tells us is that there is no logical necessity here: we could and occasionally do choose to act otherwise. For instance, many of us display *akrasia* when it comes time to book another dental appointment. Lamentable, certainly, but a very human weakness. A world which lacked *akrasia* would, I suspect, be a cleaner but a grimmer place. That said, probably 99.9% of actions which result from deliberation are designed to implement the option at the head of the agent's preference list.

To sum up, deliberation consists of the following stages:

(i) Formation of the choice set
(ii) Evaluation of the options
(iii) Decision, i.e. choice of an option.

In the following section I discuss category (iii).

3. Choice

We have followed Francine's story up to the moment she has constructed her list of preferences CAB, and is preparing to sign one of the letters of acceptance. The process of evaluation is terminated. Does this mean that there is only one of the three letters which it is possible for her to sign? That it has become, in effect, physically impossible for her to sign the A-letter or the B-letter, leaving only the C-letter as the sole open alternative? No. The deliberative model I am proposing is one in which all options of the choice set are open throughout evaluation, and remain open right up until the moment of choice. It is only the act of choice itself, the selection of a single option and the non-selection of the others, that closes all but one of the options, rendering them in effect "unchoosable" because unchosen. When the process of evaluation is terminated, but before the final choice is made, there may be a moment in which all options remain open, but one of them has not yet been selected. Or alternatively there may not; there may be no temporal gap between evaluation and choice. Nevertheless it is part of the model that there is a particular "act of choice", which may or may not coincide temporally with the end of evaluation, but which is separate and distinct from the construction of a list of preferences.

Prior to the act of choice, each option A_i belonging to the choice-set remains *open* in the sense that it is physically possible for the agent to perform whatever action initiates A_i's implementation. This requires some explanation. An action X at time t_2 is *physically possible* at time t_1 if there exists at t_1 no state of affairs which causally necessitates the non-performance of X at t_2. For example, there exists nothing at time t_0 which causally necessitates the non-occurrence of my lifting 50 kg at time t_2. But if I strain a tendon at t_1, then there does exist a state of affairs which causally necessitates my non-lifting of 50 kg at t_2. Consequently what was physically possible at t_0 is no longer physically possible at t_1. In the case of Francine each one of the options A, B, C, D remains open throughout the period of evaluation, since it is physically possible for Francine at any point to sign and mail the corresponding letter, thus beginning the implementation of the corresponding alternative. Alternatively she can tear up all three letters and re-consider her options.

It is of course true that while Francine can begin the implementation of A, B or C at any time, she cannot single-handedly ensure that the chosen option will eventuate. Unbeknowst to her, for example, the international agency administering the volunteer program in Africa may have ceased operations. Similar considerations apply to all choice sets. When we deliberate, the different alternatives are "open" to us only in the sense that we have the power to initiate the implementation of each one of them, although we do not in general have the power to ensure that any of them is finally realized. In deciding to turn on the lights I reach across to the switch, but in that short interval the electricity goes off and my action fails. What this shows is that the "open-ness" of an option during deliberation means only this: the agent has the power to *act in such a way as to begin implementation of that option*. By "act" I mean "move one's body" rather than (say) "will one's body to move". An agent who was completely paralyzed could not engage in practical deliberation, in my understanding of the term, since none of the options in her choice set would be open to her. She could engage in what I call "cognitive deliberation", but that is another story.

Consistently with the definition just given of "open-ness", the picture we have of Francine's brain during the deliberative process is this. Corresponding to the options A, B, C, D there will be neural states n(A), n(B), n(C), n(D), each of which Francine has the power to bring into existence. When and if each comes into being, it causes a bodily movement which leads to the implementation of the corresponding option. Thus neural state n(A) initiates the signing of the Sorbonne letter, n(B) the Chicago letter, etc. During deliberation and before decision, Francine's brain is in none of the states n(A), n(B), n(C), n(D), although each state is continuously available and accessible to it. I shall say more about this conception of the deliberating brain in what follows.

4. The Role of Indeterminism in Evaluation

The overall objective of this chapter is to form a conjectural empirical model of what goes on in our heads during deliberation and decision. The story of Francine provides a phenomenological description of decision-making and choice, but we are still lacking a description of the underlying physical processes in Francine's central nervous system which make the phenomenology possible. My aim is to construct, in broad outline, an empirical hypothesis as to what these neurological processes must be like. As a first step let me discuss the role that indeterminism plays in deliberation.

Philosophers who think about human action and freedom are traditionally divided into *compatibilists*, who follow Hume in considering freedom to be consistent with strict physical and neurophysiological determinism, and *libertarians*, who hold that we cannot be free unless determinism is false. Those who support libertarianism generally focus on the act of choice as the point at which any purely deterministic account of decision-making fails. Compatibilists, on the other hand, see choices as determined along with all other events in the world, and rightly point out that one cannot hold an agent X responsible for her choices unless the latter are connected with X's character. For a determinist, actions "flow from character" as water flows downhill, and a random, indeterministic event which breaks the agent/action link destroys responsibility. Worse, to the extent that an act of choice contained a random or indeterministic element, that choice would be uncontrolled by the agent whose choice it was. By definition, a determinist would say, what is due to chance falls outside anyone's power to bring about or influence. Consequently, no adequate model of responsible or controlled free will would seem to be possible if acts of choice are undetermined events.

The dispute between libertarians and compatibilists, phrased in this way, would appear to give compatibilists the upper hand. But, given our analysis of the open-ness of choice sets during deliberation, it is not difficult to see that there exists a different and more fundamental way of basing deliberation upon indeterminism, a way which preserves responsibility and control in a natural manner.

As was seen, corresponding to every choice set $\{A_i\}$ of options there is a set of mutually exclusive, non-actual neural states $n(A_1)$, $n(A_2),\ldots$, each of which is physically possible relative to the overall state of the central nervous system during deliberation. These states remain possible but non-actual up to the moment of choice, when one and only one is actualized. A physical system S with two or more qualitatively distinct non-actual future states, any one of which can be actualized and characterize S at any future moment, is *indeterministic*. A *deterministic* system has the property of possessing only one physically possible future state into which the system can move at every instant. A system which deliberates is not of this kind: the existence of a choice set of open alternatives for a deliberating subject requires indeterminism. The human central nervous system, if it is to provide the deliberator with a set of constantly accessible neural states over a continuous period of time, must be more radically indeterministic than a system composed of a sequence of chance events. The functioning of a deliberating brain is based not on individual random events identified with "choices", but on a stable indeterministic system which keeps distinct alternatives continuously available. Lacking this more sophisticated variety of indeterminism, deliberation as we know it would be impossible.

A deliberating system, then, is characterized by (a) a constantly open set of distinct non-actual states, any one of which can be actualized at any moment, and (b) the power of practical reason or practical judgement, which weighs and compares options. We have seen that condition (a), the existence of a choice set, requires indeterminism. Without indeterminism, no choice set could have more than one member. But what of the act of choice itself, which closes off all options save one? Is this act also an undetermined event?

Before addressing this question in the next section, we should note parenthetically that very few decisions are final in the sense of being irrevocable. An act of choice terminates any given deliberative process, but there is nothing to prevent the deliberator from having second thoughts and starting a new deliberation with the same choice set. People who frequently change their decisions in this way do not always endear themselves to their friends, and in bridge there are formal penalties for those who revoke a card and play something else. Nevertheless, that people should have the ability to change their minds when called for has saved individuals as well as nations from disaster on many occasions. All of which is to say that an option which is closed off by a decision at time t_1 may be re-opened by a new deliberation at t_2, and the indeterministic neural system which opens and closes options must be flexible enough to permit this.

5. The Role of Indeterminism in Choice

We turn now to the act of choice. Choice or decision makes actual one and only one of the neural states $n(A_1)$, $n(A_2)$, . . . corresponding to a choice set $\{A_1, A_2, \ldots\}$, thereby causing us to move our body so as to initiate the implementation of the corresponding option. Choice is frequently singled out by libertarians as THE indeterministic element in free action, the element which liberates us from the laws of physical necessity. But in the model of deliberation and decision which we are elaborating, a different picture of choice will emerge.

Robert Kane, in a work which explores at great length the libertarian tradition in philosophy (Kane 1996), discusses various examples of decision-making which illustrate the role played by indeterminism in what Kane calls "plural rational choices". These are rational decisions of agents faced with two or more alternative courses of action. One of his examples is that of Jane, who is deliberating whether to spend her vacation in Hawaii or Colorado (pp. 107–8). Jane consults travel books and brochures, imagines contrasting scenarios, thinks about her bank account, and eventually comes to the conclusion that all things considered, Hawaii is the best option. She therefore decides to go to Hawaii. Kane's second example is that of an engineer, a recovering alcoholic trying to save his marriage, who working late at night under great stress is tempted to have a drink to relieve the tension (pp. 126–36).

Kane's problem is how to give a coherent description of Jane's and the engineer's choices which makes them both (a) undetermined, and (b) comprehensible and rational. The difficulty is that (a) appears to negate (b), making it impossible for a choice to be both undetermined and rationally explicable in the "two-sided" or "plural" sense that Kane is seeking. As it turns out, the two choices faced by Jane and the engineer are not on a par in this respect, and will have to be treated differently.

Kane's definition of an undetermined event is one which, given the totality of antecedent conditions immediately prior to it, might have failed to occur. More fully, if D represents the state of the world immediately prior to a time *t*, then an event E is *undetermined* at *t* if D includes neither a set of conditions causally sufficient for the occurrence of E at *t*, nor a set of conditions causally sufficient for the non-occurrence of E at *t*. This implies that, if Jane's choice to vacation in Hawaii is undetermined, "*exactly the same prior deliberation* up to the moment of choice, through which she came to believe that Hawaii was, all things considered, the best option, may have issued in the choice of Colorado" (p. 107). As Kane notes, this would be strange indeed. Jane's choice of Colorado, following her painstaking evaluation and considered judgement that Hawaii was the best option, would be the opposite of rationally explicable. It would be arbitrary, random, capricious, irrational, akratic, all the things that determinists find objectionable in the concept of libertarian free choice.

What Kane is seeking is an undetermined act of choice that makes *each* option rationally explicable, i.e. supported by the agent's reasons and motives. In the case of the overworked engineer, for example, it would be understandable if, fighting temptation, he refused a drink, *and* it would also be understandable if he were overcome by desire and succumbed. "If the engineer chooses not to drink, it will be because he wanted to save his marriage and career, and if he chooses to drink it will be because of his strong (though resistable) desire to relieve tension" (p. 136). Here at last it might seem that we have a genuine example of "plural rationality", a case in which each of the two possible outcomes of an undetermined act of choice is supported by agent-reasons, and is consequently rationally explicable. But in fact there are important differences between Jane's choice and the engineer's choice which make it doubtful that the latter is a true example of "plural rationality", or indeed of deliberative choice at all.

Jane's decision of where to vacation is a typical instance of Aristotle's *prohairesis*: deliberative choice based on rational evaluation of options when all the pros and cons are open to inspection, weighed, and judged. If asked why she chose Hawaii, Jane would be able to give a meticulous account of her reasons. But the engineer's decision is not like this. The engineer (if I understand Kane's example correctly) has made a prior resolution to abstain from alcohol, and is fighting against a strong desire to have a drink. If he yields, it will not be a case of *deciding* to yield so much as *being overcome*. Now it is possible that I have misunderstood, and that Kane intended the engineer's situation to be more like Jane's, i.e. the engineer in rational fashion weighs the two alternatives of remaining abstemious to preserve his marriage vs. yielding to temptation in order to relieve psychological pressure. In that case the engineer's decision will be a "plural rational" choice, made for a reason, rather than an event of being overcome by desire or alternatively hardening one's resolve against temptation. But under this interpretation, the old incompatibility with indeterminism obtains.

If the engineer resolves for good and sufficient reasons to remain sober, then he could not select the alternative of having a drink without his choice appearing arbitrary, random, capricious, and irrational. If on the other hand, after due deliberation, he judges it would be best to have a drink and relieve the tension, then he could not, given that judgement, choose sobriety in any but a capricious manner. Given his reasoned judgement either way, his act of choice may be one-way rational, but it cannot be two-way rational.

The conclusion we are led to is that it is impossible to combine the notion of an indeterministic choice, which requires that either one of two opposed outcomes can result from a single set of initial conditions at the moment of choice, with "plural rationality". Either (i) the so-called "indeterministic act of choice" is not an act of *rational* choice at all, but instead is an example of being overcome by desire—something like the indeterministic crumbling or resistance of a concrete wall subjected to pressure. Or (ii) the act of choice is rational, as in the case of Jane's decision, but one-way rational rather than two-way rational. Like oil and water, plural rationality and undetermined acts of choice do not mix.

(What has just been said does not apply to Buridan-type choices, in which the different options have identical, equally balanced reasons for and against. In such situations an undetermined act of choice may well be rational, indeed "plural rational". But the consideration of Buridan choices must await another occasion.)

6. Deliberation as an Indeterministic Process

Does the impossibility of combining plural rationality with indeterministic choice imply the failure of libertarian analyses of deliberation and decision? Not at all. It implies merely that, if associated with individual acts of (plural) rational choice, indeterminism has, so to speak, gotten into the wrong place, i.e. been mislocated. Our purpose here is to re-position it, at the same time affirming its importance.

The culmination of Jane's deliberation about where to vacation is a decision which, relative to her final evaluation, is one-way but not two-way rational. This means that her decision of Hawaii is rational, given her evaluation, but the choice of Colorado, had it occurred following exactly the same evaluation, would have been irrational. Does this imply that no process of deliberation can be "two-way" rational? No. An indeterministic deliberative process, once begun, can result in the selection of any one of the open options of the choice set, and the selection in each case can be rational. The trick is to give a consistent description of deliberation which makes this possible.

When she begins her deliberative research into vacation plans, there is no certainty that Jane will choose Hawaii. At that early stage, we may suppose, the question of Hawaii vs. Colorado is completely undetermined. But in the course of her study and evaluation Hawaii gradually emerges as the better overall choice, the deliberation ending when Jane purchases her ticket to Honolulu. Each step in the process, we may assume, is "rational" in the sense that Jane would be able to explain to a friend, or to a philosopher doing research on decision-making, why she made it, e.g. why she preferred surfing to white-water rafting, or why she could contemplate extending her budget to include the cost of a beach-house.

Does this imply that Jane's decision is, from the start, only one-way rational? No: it is also possible that her investigation and her interests might have taken off in a different direction, or might have changed direction in the course of her deliberations. The geological history of the Grand Canyon might at a given point have caught her interest, leading her to investigate the possibility of trekking along the Canyon walls on horseback, visiting a cattle ranch, etc., all at considerably less expense than Hawaii. In this case her deliberation ends in the choice of Colorado, and again a friend to whom she explains her reasons also concludes that the choice is entirely reasonable. Consequently, when we spread the indeterminism out over the whole deliberative process, instead of concentrating it on the final act of choice, we are able to construct a model of libertarian deliberation which is both indeterministic and "plural rational".

It might be thought that the concept of an "indeterministic process", used to describe Jane's deliberation, reduces to that of a *deterministic process*, containing a set of discrete *indeterministic events*. This is not so, although individual chance events can form part of a process which is in a real sense *continuously indeterministic*. For example, in turning the pages of a travel book, Jane's fingers might by chance have skipped over an exceptionally striking photo of the Grand Canyon, which if she had seen it would have tilted her attention more in the direction of Colorado and away from Hawaii. But although the indeterministic course of Jane's deliberations is to some degree influenced by chance events like this, it is also influenced by episodes of continuous "process-like" indeterminism, in which (say) the comparative degree of weight that Jane places on budget vs. comfort in accommodation varies in a continuous though undetermined way over several hours of thought. Here there is more than one "rational" scale of comparative weighting that can be explained to a friend or observer, and the weighting influences Jane's evaluation in a gradual, prolonged way rather than as a series of discrete undetermined events.

For these reasons, Jane's deliberation is properly describable as a genuine *indeterministic process* which can take many alternative routes, each one of them capable of being defended as rational. An agent who engages in a lengthy deliberative process of this kind controls its course at all times, and is able to review, revise, or alter its direction as her judgement dictates. The process is rational throughout; each step can in principle be justified to an outside observer.

7. Conclusion

Single undetermined acts of choice, with the exception of Buridan-type choices, can at best be one-way rational, not plural rational. For this reason a libertarian account of decision-making which placed the entire burden of indeterminism on a single final "act of choice", would be unable to explain how it is that a deliberator, in the course of a single deliberation, is capable of rationally choosing an option A, and also capable of rationally choosing a different option B.

But if it is recognized that indeterminism can be predicated of *processes* as well as events, the picture changes. What is an indeterministic process? An indeterministic process operating within a system S is a process that can, at any instant, result in the coming-into-being of one of a set of distinct mutually exclusive states of S, in our case one of the neural states n(A), n(B), n(C),...corresponding to a choice set. Nothing at the start of the process, or during it, dictates *which* of the states will occur, i.e. at every instant it is causally possible that n(A) will occur, causally possible that n(B) will occur, etc. This situation persists right up until the end of the indeterministic process, when one of the states n(A), n(B), n(C),...*does* occur. At that point the process stops. It is this procedure that keeps the options in a deliberation continuously open: if the deliberation were not based on a *continuously indeterministic process*, the options would not always be open. Some such neurological model, I conjecture, must be operating in any central nervous system capable of deliberating.

Finally, the model must account for the capacity of a rational deliberator to *control* the course of her deliberation. Because she controls it, the agent is able to justify the eventual decision, whatever it may be, as free and reasons-based, not arbitrary or capricious. No less, and I believe no more, can be demanded of a consistent libertarian account of practical deliberation and decision. Plainly, the concept of "control" plays an essential part in the libertarian account outlined in this chapter. How can deliberative decision-making be a process that is at one and the same time both indeterministic and controlled? For discussion of these further issues see McCall and Lowe (2005) and McCall (2009). Not until a satisfactory account is available of a *controlled indeterministic process*, I believe, will an adequate libertarian theory of action be possible.

References

Kane, Robert (1996). *The Significance of Free Will*, New York: Oxford University Press.

McCall, Storrs and Lowe, E.J. (2005). "Indeterminist free will", *Philosophy and Phenomenological Research* 70, pp. 681–690.

McCall, Storrs (2009). *Contribution to Philosophy of Action: 5 Questions*, ed. J. Aguilar and A. Buckareff, Copenhagen: Automatic Press/VIP.

{ 5 }

Indeterminist Free Will

Storrs McCall and E.J. Lowe

One of the many reasons, I believe, why philosophy falls short of a satisfying solution to the problem of freedom is that we still cannot refer to an unflawed statement of libertarianism.

—Wiggins (1973; 33)

Wiggins's statement, made thirty years ago, remains true today. We still lack a coherent, consistent indeterminist account of how human beings deliberate, employ practical reason, make decisions, and act. But we are getting closer. Our aim in this chapter is to show that a viable description of indeterminist free will is within reach. Most of the ingredients in such an account are to be found in already-published writings, in particular Dennett (1978), (2003; Ch 4), van Inwagen (1983), Ginet (1990), Kane (1996), Nozick (1981), O'Connor (2000), Searle (2001). What is left is to put them together into a finished whole.

1. The Replay Argument

We begin by defending indeterminism in action against the charge that it leads to chance, and hence negates freedom. The defense lies in locating more precisely the position of indeterminism in the process of decision-making. This will lead in turn to an overall model of indeterministic free choice and free action.

In a paper entitled "Free will remains a mystery" (2000/2002), Peter van Inwagen reverses his earlier position in (1983) and provides arguments strengthening what has come to be known as the *Mind* argument against indeterminist free will. Van Inwagen had summarily rejected the *Mind* argument in (1983; 128–29), but returns to it in (2000/2002) and somewhat surprisingly comes to the conclusion that a new version of it, the "Replay argument", is valid after all. He presents it using the example of Alice, who in a difficult situation is faced with a choice between lying and truth-telling (2002; 169). Alice seriously considers the alternatives and freely chooses to tell the truth, although she also contemplated lying, and would have done so had she chosen. Assume the libertarian hypothesis that Alice's telling the truth was *undetermined* in the sense that immediately prior to her decision, the world contained no set of antecedent conditions which causally necessitated her action. For libertarians, Alice's act can be free only if, prior to acting, Alice is able to tell the truth *and* able to lie. This "two-sided" ability requires, and implies, indeterminism. But according to van Inwagen, if we reflect more closely on the indeterminism requirement we see that, so far from guaranteeing Alice's freedom, it in fact annuls it.

Suppose that God causes the universe to revert to its initial state shortly before Alice's truth-telling, and then allows things to go forward again. Will Alice tell the truth or lie? On the assumption that on the first occasion it was undetermined which she would do, on this occasion it will again be undetermined, and the most that can be said is that Alice *might* lie and that she *might* tell the truth. If God were to order 1000 "replays" of Alice's decision, each time re-creating exactly the same initial conditions, we could expect no more than a series of successive truth-tellings and lies on Alice's part, assuming her decision to be undetermined. Imagine that we observe (say) 493 truth-tellings and 507 lies out of 1000 trials. Will we not become convinced, as we watch this series unfold, that what will happen on the *next* replay is purely a matter of chance? Will each of the two possible decisions in each case not have an objective "ground floor" probability of about 0.5? But in that case, van Inwagen asks,

> How can we say that Alice's telling the truth was a free act? If she was faced with telling the truth and lying, and it was a mere matter of chance which of these things she did, how can we say that—and this is essential to the act's being free—she was *able* to tell the truth and *able* to lie? How could anyone be able to determine the outcome of a process if it is a matter of objective, ground-floor chance? (2002; 171)

To sum up, chance undermines free will. If replays of Alice's decision indicate a probability of about 0.5 either way, whether she lies or tells the truth in any given instance would appear to be purely a matter of chance. But in that case Alice lacks the power that one would expect of a free agent, the power to control or determine what she does. Truly, the acts of a free and responsible agent cannot be chance events.

A persuasive argument perhaps, but not a conclusive one. The weakness of the replay argument is that it puts the indeterminism of decision-making in the wrong place.

2. Indeterministic Events and Indeterministic Processes

In van Inwagen's analysis, the indeterministic element attaches to Alice's act of choice, her decision whether to lie or tell the truth. This places the whole burden of indeterminism on a single event. There is however a different model of deliberation, a model also based on indeterminism but one which predicates the indeterminism not of discrete *events*—acts of choice—but of a continuous *process*. If decision-making is analyzed in this way, as an overall indeterministic process rather than an undertaking which culminates in a single undetermined event, it becomes apparent how actions resulting from such a process can be both undetermined and free.

To illustrate the model of decision-making we have in mind, we replace van Inwagen's Alice by Robert Kane's more temporally extended example of Jane. Jane is deliberating whether to spend her vacation in Hawaii or Colorado (1996; 107–8). She takes her time, consults travel books and brochures, contemplates her bank account, and eventually comes to the conclusion that all things considered, Hawaii is the best option. At the end, she seals her decision by buying an air ticket to Honolulu. A useful way of analyzing this deliberative process (Aristotle's *bouleusis*) is to divide Jane's decision-making into three stages (McCall (1999)):

(i) *Choice-set formation* (in Jane's case identifying Hawaii and Colorado as her two options),
(ii) *Evaluation* (weighing the reasons pro and con Hawaii against the reasons pro and con Colorado),
(iii) *Choice* (Aristotle's *prohairesis*).

A necessary requirement of indeterministic decision-making is that each option in the choice-set remain open, i.e. choosable, through the entire deliberation, right up to the moment of choice. As has been frequently pointed out by both determinists (Dennett (1984; 101–122)) and non-determinists (Taylor (1964; 76)), the ability to deliberate between options A and B does not imply that A and B must each be physically possible or realizable. I may, for example, deliberate in my office about whether to walk home or take the bus, although unknown to me the buses are no longer running. A necessary condition of practical deliberation, however, is that the deliberator (1) *believe* that the options are open, and (2) be capable of initiating, by a movement of her body, the implementation of each one of them. We may imagine for example that Jane, in the final throes of her choice between Hawaii and Colorado, has placed on the table two cheques made out to two different airlines. All she has to do is sign one of them, put it in an envelope and mail it. (If the airline goes bankrupt overnight, Jane's option is nevertheless open in the sense that by signing the cheque she can *begin* its implementation, although unforeseen circumstances may later intervene to prevent its realization.)

It may not seem that replacing undetermined events by undetermined processes changes much. Could God not replay Jane's decisions as He did Alice's? In the next section it will be shown that unlike Alice-replays, Jane-replays do not sanction the inference that what Jane decides is a matter of chance.

3. Decisions Based on Reasons Are Not Due to Chance

Jane's deliberation was asserted in the previous section to be based on a continuous indeterministic process rather than a single undetermined act of choice. This implies that Jane's two options—Hawaii and Colorado—are open and available to her throughout the entire course of her deliberation. While weighing the pros and cons, she may several times swing from preferring one to preferring the other. But her swings are not arbitrary or incomprehensible. Jane is a rational person, and each time she seems to favour one side or the other, it is because she has a reason to do so.

Imagine for example that Jane concludes, after some thought, that on the whole she prefers ocean surfing to white-water rafting, and is just about to decide in favour of Hawaii when she comes across a particularly spectacular photo of the Grand Canyon in a travel book, together with a description of the complex geology of the region. Jane remembers having heard that one can ride along the rim of the canyon on horseback, and looking in her travel guide she finds several outfitting companies who conduct tours and visit cattle ranches in Colorado and Arizona. At this point she is about to change to Colorado, but when she looks at the financial side she discovers that even to rent a whole beach house in Maui is cheaper (and perhaps healthier) than two weeks of trail-riding and steak barbecues. So in the end she chooses Hawaii.

The point of all this is that every step in Jane's deliberation is rationally linked to earlier steps. If asked to justify these steps to a friend, Jane could do so. Why did she suddenly discover an interest in the geology of the Rockies? Because of a course she took years ago. Why did she finally choose beach life over trail-riding? Partly because of the thought of having to eat steak night after night. Etc. If God were to order a replay of Jane's decision, the result might be different: Jane might have judged that whatever the expense, the experience of photographing the Grand Canyon at sunrise would be worth it. But this outcome would not be due to *chance*. Whatever Jane chose, she would do so for a reason, and something that happens for a reason does not happen by chance. Consequently changing from the rather simplistic decision-making of Alice, where everything hinges on a single undetermined act of choice, to the complex reasons-based deliberation of Jane, shows that indeterminism need not negate free will. On the contrary, it makes it possible. But not every kind of indeterminism will do. In particular, merely adding other discrete undetermined events to the early stages of Alice's deliberation will not change it into a reasons-based deliberation that, like Jane's, is immune to the replay objection.

Imagine if you like a second random event E at the start of Alice's deliberation with two possible outcomes N and M such that if N occurred, Alice would think of a good reason to lie, whereas if M occurred, Alice would think of a good reason to tell the truth. Such an event would be useless, and the result would be unlike what one would expect of a deliberation based on reasons. Either Alice's final choice would accord with the reason provided by E, in which case (since E was random) the replay argument would show that whether Alice lied or told the truth was due to chance. Or, alternatively, her final choice would not depend on the outcome of E, in which case it would be a random matter whether her decision was reasons-based. The same argument applies, no matter how many individual chance events are inserted into the decision-making process. If we are seeking for an indeterministic model in which Alice's decision is based on a reason, *no matter what that decision is* (i.e. where Alice's decision is "two-way rational"), then no number of discrete random events added to an otherwise deterministic deliberative process will help. What is needed is an *indeterministic process*, which is something else again.

4. The Open-ness of Options in Deliberation

It was mentioned above, in section 2, that it suffices in deliberation that each option be *believed* to be open—that someone trying to decide whether to walk or take the bus should believe that the buses are still running—although the beliefs in question may turn out to be false. Despite this, the ability to engage in practical deliberation requires open-ness of options in the following sense.

Let the options in a deliberator's choice-set be A, B, C, . . . , and let the initial state of the deliberator's brain be N. For deliberation to take place the options must be *open*, meaning that corresponding to A, B, C, . . . there are distinct mutually exclusive motor neuronal states n(A), n(B), n(C), . . . of the brain or central nervous system, each of which is physically possible relative to N at all times during the deliberative process, but none of which is actual. If it becomes actual, the state n(A) causes a bodily movement which initiates the implementation of option A, and similarly for the states n(B), n(C), etc. This "minimal open-ness" of options is consistent with the possibility that the deliberator's beliefs about the ultimate realizability of A, B, C, . . . are all false: the buses may not be running, and the options of walking home, telephoning a cab etc. may all be unavailable. Nevertheless, deliberation is possible, as long as the group of neuronal states which initiates or sets in motion each of these options is physically accessible. In this minimal sense, then, every deliberation presupposes an open available set of possible but non-actual motor neuron states corresponding to the different options. Each of these remains continuously accessible or available to the brain at all times during deliberation.

This is part, though not all, of what is meant in saying that deliberation is a continuous indeterministic process. The indeterminism of free choice lies not only in the unpredictability of decisions, but more importantly in the ability of the deliberating intentional system to maintain at least two open options during the deliberative process. A deterministic system cannot do this: the maximum number of alternatives it can keep open is one.

5. The Weighing of Options: A Controlled Indeterministic Process

When Jane deliberates, she weighs the various reasons for and against Hawaii with the various reasons for and against Colorado. If she is methodical, she may make a list:

Hawaii	For:	Relaxing beach life
	Against:	Costly air fares
Colorado	For:	Breathtaking scenery
	Against:	Saddle sores and steaks

Every deliberation process contains a similar phase, named "Evaluation" in section 2, which involves the weighing of reasons pro and con the options.

But what sort of weighing is this? Is it like weighing sugar? Is the weighing process deterministic or indeterministic? Is it partially or wholly under the control of the weigher? If so, by what method could the weigher, i.e. the deliberator, influence the outcome of the weighing process? How is it possible for such a process to be both controlled and indeterministic?

To make sense of the idea of controlled, indeterministic weighing is not easy. But it must be possible to do so, because we all engage in this sort of weighing every time we plan a vacation, evaluate a dossier, or visit a supermarket. The example of Jane provides a good case study, and we shall continue to use her as our model.

When Jane deliberates, she not only *weighs* the reasons for and against Hawaii and Colorado, she *weights* them. Reasons like "relaxing beach life" and "breathtaking scenery" do not come with ready-made weights, which Jane balances one against another as one might balance a package of sugar on a scale. On the contrary, before she compares the weight or degree of importance of one reason with that of another, Jane must *assign* a weight or degree of importance to each reason. Robert Nozick, in his discussion of deliberation and choice in *Philosophical Explanations*, makes this point:

> The reasons [considered in deliberation] do not come with previously given precisely specified weights; the decision process is not one of discovering such precise weights but of assigning them. (1981; 294)

Unfortunately, Nozick postpones the time of assigning weights to reasons in deliberation to the moment of decision itself. In his words, "the weights of reasons are inchoate until the decision.... A decision establishes inequalities in weight, even if not precise weights." But in so doing, Nozick removes the possibility that Jane's decision should, even in part, be dependent on, or be explained by, previously assigned weights. In his account, these weights are established only *after* the decision, and therefore cannot be part of the sequence of events which led up to it. As is made clear in O'Connor (2000; 30–32), Nozick's *post hoc* method of assigning weights to reasons cannot shed light on how Jane is able to exercise a degree of control over her decision *before* the decision is made.

The key question is, how is Jane able to regulate the weights assigned to the various factors of *beach house comfort, breathtaking views, stubborn trail horses*, etc., all of which enter into and compete with one another in the evaluation process? What tilts the balance? The answer lies in Jane's character: she is a rational deliberator, someone who uses her judgement. We may imagine an interior dialogue going on in Jane's head: "Why attach so much importance to a beach house? It's comfortable and informal, granted. But surely very expensive? Yes, but what am I saving my money for anyway? Isn't this my only vacation in two years?" This is not the dialogue of a deliberator pulled this way and that by conflicting desires, aversions and emotions. It is more like the internal dialogue of a judge, who in writing up an opinion pauses over each step, conscious of the danger of being over-ruled on appeal.

Contrary to what Hume says, reason is not in the judge's case the slave of the passions, but is the exquisite tool with which he shapes his decision. Sometimes, perhaps most times, when judges sit down to write their decisions, they know from the start which way the decision will go. But other times they genuinely don't know: they use their active reason to discover the steps and links which lead them to a decision, somewhat like a logician setting out to prove a difficult lemma before he knows whether the lemma is true.[1] So it is, we claim, with Jane. Before she deliberates, she doesn't know whether it will be Hawaii or Colorado. She employs her reason to find out, using her judgement to adjudicate between the competing claims, the clash of pros and cons. At all times, the process is strictly under her control.

We are now able to square the circle, to see how deliberative processes in which the deliberator uses her power of judgement can be (i) *indeterministic*,

[1] We owe this glimpse into a judge's life to Peter Heerey. Hodgson (1991; 136–141) contains interesting material on the role of judgement and weighing in legal reasoning. Another example of a controlled, indeterministic, reasons-based process is Andrew Wiles's 20-year effort to prove Fermat's Last Theorem (Singh 1997). At no point until the end was it definite that he would succeed or definite that he would fail, and throughout the 20 years the process remained supremely rational and controlled by Wiles.

(ii) *reasons-based*, and (iii) *controlled*. They are (i) indeterministic because the eventual outcome is not fixed by the initial conditions. If God were to replay Jane's deliberation 1000 times, the result might be 452 Hawaii-decisions and 548 Colorado-decisions. But (ii) in no case are these chance decisions: each one is based on reasons and the path that leads to it is rational and justifiable. Finally (iii) the entire process is controlled by Jane, who uses her judgement to weigh and weight the different reasons. Since the weighing and weighting are judgemental, nothing guarantees that identical weights will be attached to, say, "beach house living" on two separate occasions, if God were to order a replay. It is, typically, the free exercise of Jane's power of judgement which makes reasons-based deliberation indeterministic, not the occurrence of random or unpredictable events. Jane's decision-making is controlled, undetermined, and "two-way rational", meaning that whatever her decision is, it will be based on reasons.

It is now possible to attach a more definite meaning to the notion of an indeterministic *process*, as distinct from an indeterministic event. An indeterministic event E, e.g. an event occasioned by the flip of a coin or the roll of a die, is characterised by the sudden chance occurrence of one of a set of two or more possible events or outcomes, relative to given initial conditions. At the time of E, the probabilities of the outcomes all jump instantaneously and discontinuously from some positive value between 1 and 0 to a value which is either 1 or 0, depending on whether the outcome is realized. In an indeterministic process on the other hand there are also possible outcomes, but instead of jumping the probabilities of the outcomes evolve smoothly and continuously throughout the process until the last moment (the "decision"), when they change smoothly into 1 or 0. An indeterministic process of the kind we have referred to as "reasons based" may exhibit dramatic probability-swings, but these swings will be perfectly "rational" or understandable because associated with weighted reasons. The weighting is provided in a controlled way, by the deliberator's judgement.

6. The Consistency of Libertarianism

Wiggins's challenge, set thirty years ago, was to give a consistent description of libertarian freedom. This we believe we have done. Libertarianism is a consistent idea if there exists at least one species of human behaviour which is free, rational, controlled by the agent, and indeterministic. Rather than attempt to cover many different kinds of rational activity we have focussed on only one, namely action consequent upon a rational, controlled deliberative process in which the agent is faced with at least two possible courses of action and freely decides on one of them. There may be other exemplifications of libertarian free will, but to establish consistency does not require them.

The main features of the indeterministic deliberative process which demonstrates consistency are as follows.

(1) An agent X is faced with deciding between options A, B, C, ...
(2) There are, in X's estimation, reasons for and reasons against each option.
(3) X uses her power of rational judgement to weight these reasons and to weigh one option against another.
(4) The process of weighing and weighting is controlled by X's judgement, is on-going throughout the deliberation, and is justifiable to a third party.
(5) Each option remains open (choosable) up to the moment of decision.
(6) The deliberation ends with X's reasoned choice of one of the options.

Conclusion: Rational, indeterministic, controlled deliberative processes prove that the concept of libertarian free will is internally consistent.

References

Dennett, D.C. (1978) "On giving the libertarians what they say they want", in *Brainstorms*, Cambridge, MA: Bradford Books.

—— (1984) *Elbow Room*, Cambridge, MA: MIT Press.

—— (2003) *Freedom Evolves*, New York: Viking Press.

Ginet, C. (1990) *On Action*, Cambridge: Cambridge University Press.

Hodgson, D. (1991) *The Mind Matters*, Oxford and New York: Oxford University Press.

Kane, R. (1996) *The Significance of Free Will*, New York: Oxford University Press.

McCall, S. (1999) "Deliberation reasons and explanation reasons", in Jackendoff et al. (eds) *Language, Logic, and Concepts*, Cambridge, MA: MIT Press.

Nozick, R. (1981) *Philosophical Explanations*, Cambridge, MA: Harvard University Press.

O'Connor, T. (2000) *Persons and Causes*, New York: Oxford University Press.

Searle, J. (2001) *Rationality in Action*, Cambridge, MA: MIT Press.

Singh, S. (1997) *Fermat's Last Theorem*, London: Fourth Estate.

Taylor, R. (1964) "Deliberation and foreknowledge", *American Philosophical Quarterly* 1, 73–80.

van Inwagen, P. (1983) *An Essay on Free Will*, Oxford and New York: Oxford University Press.

—— (2000/2002) "Free will remains a mystery", originally published 2000, reprinted in R. Kane (ed.) (2002) *The Oxford Handbook of Free Will*, pp. 158–177. Oxford and New York: Oxford University Press.

Wiggins, D. (1973) "Towards a reasonable libertarianism", in T. Honderich (ed.) *Essays on Freedom of Action*, London: Routledge and Kegan Paul.

{ 6 }

The Determinists Have Run Out of Luck—For a Good Reason

Storrs McCall and E.J. Lowe

In his paper "Bad luck once again" Neil Levy attacks our proof of the consistency of libertarianism by reiterating a time-worn compatibilist complaint.[1] This criticism is that what is not determined must be due to chance. If A has a choice of X or Y, neither X nor Y being causally determined, then if A chooses X it can only be by *chance*, never *for a reason.* The only "reason" that could explain the choice of X over Y would have to be a *causally sufficient reason*, which would rule out A's having a genuine choice in the first place. Either X is causally necessitated or X is realized by sheer luck. But that these are the only alternatives is untrue. The exercise of deliberative reason opens the way between the Scylla of causal necessitation and the Charybdis of chance, as we shall try to make clear.

The central core of Levy's argument is that any attempt to give a reasons-based explanation of a contrastive fact must fail. A contrastive fact is a fact of the kind "Jane decides to vacation in Hawaii rather than Colorado," or "Jane assigns a greater weight to surfing than to white-water rafting." In the last three paragraphs of his paper Levy argues that Jane's assigning more weight to surfing than to rafting cannot be a reasons-based assignment, because, as he puts it, "the reasons that would explain the weighting are the weighted reasons *themselves*." Similarly, prior to making her Hawaii/Colorado decision, Jane has reasons for both options, but no *decisive* reasons for either. As he says, "Her reasons have got her this far, but they can take her no further." Lacking a decisive reason, by which Levy

---

[1] Neil Levy, "Bad Luck Once Again" (next article), discussing McCall and Lowe, "Indeterminist Free Will," *Philosophy and Phenomenological Research* 70 (2005), pp. 681–690.

apparently means a reason causally sufficient to produce Jane's choice, her decision must be due to pure chance or luck.

Levy's arguments are fallacious. Clearly, for him, a reason for A's doing or choosing to do X cannot count as "decisive" unless A's having that reason necessitates her doing X. It "explains" her action simply by excluding the possibility of any other course of action. But a libertarian cannot accept such a notion of a "decisive reason." For libertarians, A's reason for doing X was "decisive" if and only if it was, *in fact*, the reason for which A did X. The "decisive" reasons are just those that the agent *did* act upon, as opposed to those she could have acted upon but did not. Reasons which do not necessitate can still explain why this course of action resulted rather than that, contrary to Levy's apparent assumption. And such reasons are absent altogether in genuine cases of luck or chance.

Levy suggests that our position is no better than Kane's. But we suggest on the other hand that Levy's objection is no better against Kane's position than it is against ours: it simply begs the question at issue between the libertarians and their opponents. Levy has failed to perceive what is distinctive about our position and makes it superior to Kane's, namely that the evaluation stage of deliberation is a continuous indeterministic process rather than a punctuated series of chance events. Despite what he says, the weighting of reasons can be reasons-based, and a decision to act on the reasons for X rather than the reasons for Y can be, and frequently is, rationally explicable and defensible. This emerges clearly in the example of Jane's vacation deliberations.

Suppose that Jane asks Susan, a friend, to help her make up her mind. She reviews with Susan the pros and cons of Hawaii versus Colorado, the luxury of a beach house, the incredible sunsets over the Grand Canyon, the danger of trail horses stumbling, the expense of flying halfway across the Pacific, etc. Over a period of days, during which time the two options remain continuously open and choosable, Jane attempts to make precise the importance she attaches to these different reasons pro and con. It turns out, perhaps, that Jane has an exaggerated fear of rattlesnakes, loves the smell of campfires, adores snorkeling but has never forgotten the horror of the movie "Jaws," etc. With Susan's help, she eventually arrives at a scale of weights attached to every conceivable feature of the two alternative vacations, and is able to articulate the reasons upon which her weighting system is constructed. Finally, on the day of decision, she makes her controlled, considered, reasons-based choice. Luck, or chance, does not enter into it. Jane's decision is the antithesis of tossing a coin. It is fully rational and yet, both options continue to be physically possible future alternative courses of action up until the moment Jane phones her travel agent. If God were to order a re-play of Jane's deliberation it is conceivable that Jane's weighting system might be slightly different, and that her choice could go the other way. A situation of this kind is describable as "two-way rational."

The difficulty that some philosophers experience in grasping the possibility of controlled, indeterministic, rational, open-ended processes which typically constitute deliberations may be explained by their neglect of the concept of probabilistic causation. Although Humean causes are invariably followed by their effects, probabilistic causes are not. It is possible that during deliberation, the activity of the central nervous system is best described as an indeterministic neurological mechanism subject to the force of probabilistic causes, these causes corresponding to the various reasons and weightings of reasons that deliberators like Jane deal with. This is not to say that reasons *are* causes, but only to suggest that probabilistic causation may be a useful concept when studying deliberative processes. The notion of probabilistic causation completely undercuts the false Scylla/Charybdis dilemma of the determinists, to which Levy appeals.

When a deliberator is deliberating, the probabilities of each option being selected vary continuously, with swings toward one option or another as new considerations occur to the deliberator, or as she attempts to justify her weighting system to a third party. Finally, with the decision, each option's probability moves smoothly and continuously to assume the value of zero or one. In this model of libertarian deliberation, there is room for probability, and for probabilistic causation, but no room for unadulterated chance or luck.

A final example, taken from judicial deliberation, shows that in some cases it may prove not just mistaken, but even risky, for a determinist to assert that a person's choice of option X rather than option Y is due to chance or luck. In June 2005 the Supreme Court of Canada brought down a decision in the case of Chaoulli vs. the Attorney General of Quebec which surprised many jurists. Chaoulli, a physician, had a patient with a number of health problems who was deprived of the timely care he needed by the long waiting times for hospital treatment. Chaoulli had tried unsuccessfully over a period of years to have his home-delivered medical services recognized and to operate an independent private hospital, for which patients could purchase private health insurance in addition to the public insurance that every Canadian possesses automatically. The Quebec government had refused to permit such private insurance on the grounds that it violated the Canada Health Insurance Act. However, Chaoulli argued that the lengthy waiting times for certain surgical procedures, which inflicted needless pain and suffering on patients and in some cases increased their risk of mortality, infringed their rights as guaranteed in the Canadian Charter of Human Rights. Section 7 of the Charter states that every citizen has the right to life, liberty and security of the person. Chaoulli argued that patients who were deprived of urgent health care when there was a remedy through private insurance were being deprived of a fundamental Charter right, and to the great surprise of many Canadians the court agreed with him. The Canada Health Act which guarantees health care to all regardless of their ability to pay is a cornerstone of Canadian society, and the Chaoulli decision, which may now force the government to permit access to some forms of health care through private insurance, is a landmark decision. Canada, it seems, is the only OECD country which prohibits insurance for health care provided by non-participating physicians.

The Chaoulli decision is an example of a controlled, rational, indeterministic process. Its relevance to the libertarian/determinist debate lies in (a) the unpredictability of, and surprise occasioned by, the Supreme Court judgment, and (b) the great care which the participating judges took in comparing the legal weight of the Canada Health Act with that of individual rights enshrined in the Charter. No doubt the eventual decision arrived at could have gone the other way. But for Neil Levy or anyone else to assert that the Supreme Court's judgement was arrived at "by luck," or "by chance," would be a bold and risky move. To stand up and say this in a legal tribunal would be to invite the charge of contempt of court. In throwing around the word "luck," determinists should be careful.

Summing up, there are in deliberation many contrastive facts which admit of reasons-based explanations. The Supreme Court decision to support individual Charter rights over the Canada Health Act is one of them, and the reasons for and against this decision will surely be argued for years to come. To say, with the compatibilists, that the decision was arrived at by luck adds nothing to the debate but obfuscation and confusion.

References

Levy, N. (2008) "Bad Luck Once Again", *Philosophy and Phenomenological Research* 77, pp. 749–754.

McCall, S. and Lowe, E.J. (2005) "Indeterminist Free Will," *Philosophy and Phenomenological Research* 70, pp. 681–690.

{ 7 }

Action Based on Deliberation and Decision

Causal theories of action aim to show that actions invariably have causes, whether these be neurological states of the agent or "primary reasons" consisting of desire/belief pairs (Davidson (1963, reprinted in 1980), Goldman (1970, pp. 72–121), Aguilar and Bukareff (2010)). Agent causation theories say that actions are caused directly by agents (Chisholm (1966), Taylor (1966), Clarke (1993), O'Connor (2000)). This chapter presents a contrasting theory in which voluntary, intentional actions result from decisions, and decisions in turn are the culmination of a deliberative process that is not a causal one. (Involuntary actions, such as sneezing or shuddering, don't result from decisions and aren't dealt with here.)

Practical deliberation (Aristotle's *bouleusis*) is the process of coming to a decision (*prohairesis*). A decision consists of choosing which one of a number of different alternative courses of action to perform. In the deliberative process there are reasons for and against each option, which are considered and weighed by the agent. But none of these reasons are causes. In the end, the agent decides on one of the alternative action-options and performs it, but her decision, though reasons-based, has no cause. In fact, I argue below for the stronger conclusion that the decision which terminates the deliberation *cannot* be caused, and that the entire deliberative process is a non-causal one. The decision causes the action, but is not itself caused. In addition, as will be shown, deliberation provides the basis for a clear distinction between *indeterminism*, on the one hand, and *chance* and *luck*, on the other.

1. Deliberation

There are different varieties of deliberation, and the type focused upon here is (i) *practical deliberation*, deliberation about what to do. In addition to practical deliberation there is also (ii) *cognitive deliberation*, deliberation about what is true and what is false, and (iii) *value deliberation*, which is deliberation about what values or goals to adopt. In a well-known passage, Aristotle says that we deliberate not about ends but about means (*Nicomachean Ethics*, 1112b13). Thus a doctor deliberates not about whether to heal a patient, but about how to heal him. This is certainly true, but in a different mode a doctor could also deliberate about whether we should seek to eliminate influenza, or whether influenza might not play an important role in stimulating the body's immune system and in so doing making us ultimately healthier. A doctor who reasoned along these lines would be engaging in (iii) value deliberation.

Finally, we engage in (ii) cognitive deliberation when we ask ourselves, for example, whether Mallory and Irving ever succeeded in reaching the summit of Mount Everest in 1924, or whether it is true or false that every even number is the sum of two primes, or whether Germany or France is the most influential nation in Europe. A nice case of the difference between practical and cognitive deliberation is the division of labour in jury trials. The jury deliberates over the cognitive question of guilt or innocence, and the judge, once a verdict of guilt has been handed down, engages in the practical deliberation of what penalty to impose. A more lighthearted example: in Canada, Parliament deliberates over (iii) how the public's money *ought* to be spent, the Auditor-General deliberates over (ii) how it *was* spent, and the Cabinet deliberates over (i) *how to spend it*.

Practical deliberation is a conscious, rational process divided into three parts:

(1) Formation of the "choice set", composed of the different alternative actions open to the agent at a given time,
(2) Evaluation of the alternatives,
(3) Decision.

Of these, the most time-consuming is (2). Some deliberations can last for hours, or even months. Consider Michael, who is graduating from university and trying to make up his mind between, (A) applying to do graduate work in microbiology, (B) getting a job in a local consulting firm, and (C) taking a year off to bicycle through Europe. Michael weighs these options, considering their pros and cons. His ambition is to do research in the life sciences, but his grades to date have been only average. His father is putting pressure on him to become employed. But perhaps a year abroad, though expensive, would enable him to fix his priorities. Each of the options A, B and C has reasons in favor of it, and reasons against it, which I refer to as "delib-reasons" (McCall (1999)). Michael deliberates. Deadlines are approaching. Finally he sits down to write a letter, either to the department of microbiology (option A), or to the consulting firm (B), or to his father explaining the need for a year off (C). One can imagine different possible motor neuronal impulses n(A), n(B) and n(C) in Michael's brain, leading to letter-writing movements of his hand and fingers which initiate and eventually actualize the different options A, B and C respectively. Michael decides, and his decision, and the subsequent bodily movements that it causes, ends the deliberation.

2. The Causal Theory of Action

Are reasons causes? More specifically, when someone decides to perform an action X, is the reason for which he does X the *cause* of his doing X? In his paper "Actions, reasons and causes" (1963), Donald Davidson created single-handedly the modern event-causal theory of action. Davidson first identifies a desire/belief pair as the "primary reason" why an agent performed an action, and then asserts that the primary reason for an action is its cause (1980, p. 4). Since Davidson holds that where there is causality, there must be a law, and since he believes there are no strict psychophysical laws, he proposes his theory of "anomalous monism". According to this theory, a primary reason under its *physical* description as a neurophysiological brain-event is the cause of a *physical* movement of the body, the latter constituting the action in question. For Davidson, reasons under the description of physical events are the causes of physical actions. But as will be seen there are good reasons to reject Davidson's causal theory (the word "reasons" being used here in a non-Davidson sense).

An agent may have several reasons for doing something, and yet end up doing it for only one of those reasons. I may want to mow my lawn early tomorrow morning to impress my visiting in-laws with the sight of a well-kept lawn. And I may also want to repay my neighbor for waking me up at dawn the other day with the roar of her power mower (see Mele (2003), p. 39). It might seem that a Davidsonian analysis of one and only one of these reasons as the *cause* of my mowing provides the best hope of singling out the real reason for which I actually mowed it. It would seem natural to assume that if reasons are causes, the strongest reason is the strongest cause (see Davidson (1980), p. xii). Alternatively, we might regard multiple alternative reasons as a case of causal over-determination, in which two or more Davidsonian sufficient causes would produce one and the same action. But in that case Davidson's analysis fails to answer the question, which of my original reasons was *the* reason for my action? I suspect that the Davidsonian causal theory cannot solve this problem. Let us return to the deliberation-decision scenario, in search of a better understanding.

As was seen above, the first step in any practical deliberation is the formation of the "choice set" of different alternative actions A, B, C, ... open to the agent at a given time.[1] The minimum number of such alternative options is two, since in the presence of only one choosable option there can be no process of deliberation. One simply acts. Such circumstances are rare, and probably in

[1] Cf. Carl Ginet: "I have freedom of action at a given moment if more than one alternative action is then *open to me*. Two or more actions are *alternatives* if it is logically impossible for me to do more than one of them at the same time. Two or more alternatives are *open to me* at a given moment if which of them I do next is entirely up to my choice at that moment." (Ginet (1990), p. 90)

99% of the situations in which we find ourselves throughout the day, we have a choice of different possible things to do. I am presently writing words on paper. Do I use a pen or a pencil? The choice is mine.

Throughout any deliberative process, all choosable alternative options A, B, C,... specific to that process remain open continuously. An "open" alternative action option is one that it is physically possible for the agent perform. It follows that deliberation is an *indeterministic process*, meaning that relative to the initial conditions prevailing at its start, two or more different terminations of the process are possible. By contrast, a *deterministic* process is one with only a single physically possible ending, where "physically possible" means "permitted by the laws of nature, relative to given initial conditions". Deliberation is a process that is necessarily indeterministic, since every such process has more than one different possible outcome.

It is worthwhile reflecting for a moment on the concept of an indeterministic *process*, as contrasted with an indeterministic *event*. Normally when we think of indeterminism we think of indeterministic events, such as the fall of a coin, the outcome of rolling a pair of dice, the decay of a radioactive particle, or the spin measurement of an electron in a Stern Gerlach apparatus. Strictly speaking, however, each of these indeterministic events is the culmination of a long or short randomizing process, such as rolling dice or simply waiting for a particle to decay spontaneously. Like the process of deliberation, such processes have specific initial conditions, relative to which there are two or more physically possible outcomes. (The existence of different possible outcomes is what makes them indeterministic; processes with only one physically possible outcome are deterministic.) Deliberation differs, however, from other indeterministic processes like tossing a coin in that its eventual outcome, the decision that terminates it, is not a chance event. There is no reason why a microphysical particle decays at a certain moment, but there is a reason why an agent arrives at the decision he or she makes. However, as will be seen, that decision has no cause.

3. Why Deliberative Decisions Have No Causes

In any deliberation, all the options facing the deliberator remain open, choosable, and physically possible up until the moment of choice. Deliberation is a conscious process, a sequence of subjective phenomenal states, every step of which is a product of the reflective intelligence of the deliberator and can in principle be described in detail to a third party. The heart of deliberation is the frequently lengthy process of assigning weights to the different options, and then of weighing one option against another (the root of the word "deliberation" is the Latin *libra*, a balance-scale). Weighing one option against another is not like weighing coffee or tea; the weights must be *assigned* to the options before any comparative weighing takes place:

> The reasons [considered in deliberation] do not come with precisely specified weights; the decision process is not one of discovering such precise weights but of assigning them. (Nozick (1981), p. 294)

Moreover, the process of assigning weights is essentially a subjective *first-person* activity, rather than one describable objectively in third-person terms. The difference between first- and third-person descriptions is related to the metaphysical distinction between appearance and reality. Thus Searle says that "where consciousness is concerned the appearance is the reality" (Searle (2004), p. 85). If it consciously *seems* to me that I am in pain, then I *am* in pain, and if in deliberation I seem to attach more weight to the option of reading than to watching TV, then I really do weight reading more heavily than TV watching.

The purpose of emphasizing the subjective character of deliberation is to underline its status as a conscious, controlled, indeterministic process, one in which earlier steps are not related to later ones as cause-and-effect. To be sure, conscious thoughts can be regarded as supervening on, and caused by, neurophysiological processes. But not even the entire neural process that underlies a deliberation can be a sufficient cause of the conscious decision that terminates it. As emphasized above, every deliberative process has two or more distinct possible outcomes corresponding to the alternative options laid down at its beginning. In establishing *which* of these outcomes becomes actual, the direction of causality runs the other way, from conscious choice to motor neuronal firings rather than from neurophysiological state to conscious choice. The reason why, in deliberation, conscious choice cannot be caused by prior neurophysiological states is the fact of the open-ness of those alternative states up until the very moment t_0 of choice. If they all truly remain open, distinct, and physically possible up until t_0, they cannot cause or determine the conscious choice *at* t_0. Nor can there exist, prior to t_0, another, distinct neurophysiological state S which causes the deliberative choice. If such a state did exist, the other alternative options forming the basis of the deliberation would no longer be open at t_0. The only conclusion consistent with these factors is that there exists a conscious, subjective choice CH that activates a motor neuronal impulse n(A), n(B), or n(C) corresponding to *one* of the different options A, B, or C. But CH itself supervenes on no neurophysiological state. It causes neurophysiological events, but is not caused by them.

4. Rational, Controlled, Indeterministic Processes

Deliberation is but one example of processes that are rational, controlled, and indeterministic, and many other examples exist. A typical one is walking down a crowded sidewalk. Other pedestrians come at you unexpectedly from different directions, and yet collisions rarely occur. Skiing is another case, in which no turn is 100% deterministic because of slight irregularities in the snow, but in which there is a world of difference between skiing in control and skiing out of control (McCall, forthcoming). Chess is a supremely rational pursuit, requiring a high degree of control, but the fact that the opponent's response to each move is unknown and often unanticipated makes every game an indeterministic process.

"Control" is a difficult concept to analyze. Fischer (1994) and Fischer and Ravizza (1998) distinguish between "guidance" and "regulative" control, the former being exercised by the driver of a car that turns right every time the wheel is turned right, but that occasionally turns right without the wheel being turned right. In this case, turning the wheel right is only a sufficient condition of the car's turning right, not a necessary condition (Fischer, pp. 132–4, Fischer and Ravizza, pp. 30–32). "Regulative" control, on the other hand, obtains where turning the wheel to the right is both a sufficient and a necessary condition of having the car turn right. This being said, the variety of control we exercise over our deliberations is regulative control, which is the control we normally have over the movement of our limbs. (In the field of thought, on the other hand, we may in fact exercise only "guidance" control, since although we can think whatever we want, sometimes ideas enter our heads unbidden.) However, acts of deliberative choice demand the use of full regulative control, in which choice is a causally necessary and sufficient condition of action, within the context of an overall indeterministic process. A good example of full, regulative control would be the proving of Fermat's Last Theorem in 1993, the culmination of 20 years of work by Andrew Wiles. At no point in this process until the end was it definite that Wiles would ultimately succeed or ultimately fail: both possibilities remained open. Over the full 20 years of his research, Wiles's project was a supreme example of a conscious, rational, controlled, indeterministic process. See Singh (1997). More will be said about control below, in section 6.

5. Deliberation Reasons and Explanation Reasons

For philosophers who follow Davidson, reasons for actions are their causes. But in deliberation, reasons play a different role. In the course of the process of evaluation there normally emerge both reasons for, and reasons against, each alternative action-option. I call these "delib-reasons". Since in any deliberation there are always two or more different, mutually exclusive options, with delib-reasons for each, a delib-reason cannot be the cause of the action eventually decided upon. In addition to delib-reasons, to be sure, there *is* a reason why, say, option B is chosen over options A and C. This reason cannot itself be a delib-reason; rather it is the reason why one set of delib-reasons prevails over another set. Let us call it an "explanation reason". Delib-reasons are reasons *to* X, or reasons *for* doing X, while explanation-reasons are reasons *why* or *that* X was done rather than Y. But explanation-reasons cannot be causes any more than delib-reasons can. There cannot be an explanation-reason for why action X was performed unless X has actually been performed. Explanation reasons, if they exist at all, come into existence only after the things for which they are explanations. Since causes necessarily precede their effects, explanation reasons cannot be causes of X any more than delib-reasons can be causes of X. Where actions are the end-result of a deliberative process, reasons cannot be their causes.

6. The Causing of a Bodily Movement by a Conscious Event

Deliberation is a conscious process, and every step in the process is a mental event. Underlying the sequence of mental events there is, of course, a vast complex of neurophysiological events, upon which the mental events can be regarded as supervening. But as argued in section 3 above neither the conscious sequence nor the neurophysiological complex is a deterministic causal one. This follows from the fact that every deliberational process has two or more physically possible outcomes.

Furthermore deliberation, though indeterministic, is a *controlled* process, and what controls it are the thoughts, likings, purposes, and reasoning powers of the deliberator. A good example of deliberational control is found in the evaluative weighting and comparative weighing of the different options open to the agent (section 3 above). The culmination of deliberation is decision, and here for the first time a deterministic, causal link appears in human agency. In the case where no temporal gap occurs between deciding and doing, a conscious decision causes a motor neuronal impulse that leads to a movement of the body. If there is a temporal gap, the decision is not a proximal cause of the bodily movement, but either causes it at a later time or leaves it open for further changes:

And indeed there will be time.... time yet for a hundred indecisions,
And for a hundred visions and revisions,
Before the taking of a toast and tea. (T.S. Eliot, *The Love Song of J. Alfred Prufrock*)

Sooner or later, every deliberation ends in a decision, even if the decision is simply to do nothing, and let events take their course. Whether the outcome is action or inaction, bodily movement or no bodily movement, the end result of a deliberation is "up to" the deliberator. He or she controls the process, including the outcome, and deliberation-resulting-in-decision is an *autonomous* process, not a heteronomous one. It is a prime example of mental causation, the determining of future events by conscious choices.

7. Indeterminism, Chance, and Luck

Perhaps the most serious objection to the model of action built on deliberation and decision presented here derives from the essentially *indeterministic* character of the deliberative process. Without multiple alternative action-options, there can be no such thing as deliberation, and no such thing as the decision that ends it. But for many it is difficult to believe that if deciding is the result of the genuinely indeterministic process, then deciding to do X, and doing X, are not matters of sheer chance or luck. As J.J.C. Smart put it many years ago, every event is either causally determined, or, if not, is due to chance (Smart (1961), p. 296). But is this so? Is it possible to find certain types of actions, those resulting from deliberation, that are rooted in indeterminism without hanging upon luck? Is there room to pass between the Scylla of determinism and Charybdis of chance? I shall argue that there is.

The issue of chance vs. determinism in the theory of action has been debated by many philosophers over many years, most notably R.E. Hobart (1934), A.J. Ayer (1946), reprinted in (1954), J.J.C. Smart (1961), Robert Nozick (1981) Ch. 4, Peter van Inwagen (1983), Galen Strawson (1986), Randolph Clarke (1995), (2003), Robert Kane (1999), (2005), Al Mele (2005), (2006), T. Pink (2004), McCall and E.J. Lowe (2005). But the discussion to date has tended, too easily, to identify "undetermined" events with "chance" events. Moving from events to processes, we can see that not every outcome of an indeterministic process need be a random or chance event. If the process in question is both indeterministic and controlled, its outcome may be planned, deliberate and intentional: the antithesis of chance. Examples of such processes were discussed above in section 4. When I walk down a crowded street, it's not a matter of *luck* that I don't bump into people, even though the trajectories of other walkers may be unpredictable and indeterministic. Sinking a putt in golf is *partly* controlled, partly luck, depending on irregularities in the green. So control is a matter of degree. But walking down a crowded sidewalk is, for most people, *fully controlled*, as is playing chess. One would not say that a chess player lost a game because of *bad luck*, but because of *bad play*.

On the other hand, it *is* a matter of luck when I roll seven with two dice, or get heads when I toss a coin, or happen to detect the decay of a microparticle at a certain time. These processes are uncontrolled. Controlled processes can still be indeterministic in the sense of having different possible outcomes, but the controller is responsible for which of the possible outcomes becomes actual. If I deliberate and decide to go to a movie at 7 pm, it's not by luck or chance that I make that decision, even though I could have decided to stay at home.

Al Mele has given careful thought to the following question: how is it that someone is able to decide to perform an action A on one occasion, while in other exactly similar circumstances he decides not to perform A? "In the actual world, Joe decides at *t* to *A*. In another world with the same laws of nature and the same past, he decides at *t* not to *A*. If there is nothing about Joe's powers, capacities, states of mind, moral character, and the like in either world that accounts for this difference, then the difference seems to be just a matter of luck. And given that neither world diverges from the other in any respect before *t*, there is no difference at all in Joe in these two worlds that can account for the difference in his decisions. To be sure, something about Joe may explain why it is *possible* for him to decide to *A* in the actual world and decide not to *A* in another world with the same laws and past. That he is an indeterministic decision maker may explain this. That is entirely consistent with the difference in his decisions being just a matter of luck." (Mele (2005), p. 384).

These considerations have led Mele to propose a tentative model of what might go on inside the heads of deliberators faced with deciding whether or not to perform an action A. Assume that both options are open to them and that there is, say, a probability p of their deciding to do A and $1-p$ of deciding not to do A. In Mele's model, a tiny neural roulette wheel is divided into segments, with proportions p and $1-p$ corresponding to the probabilities of the two options. A little neural ball bounces along the wheel, and its landing at a particular segment constitutes the agent's making the corresponding decision (Mele (2005), p. 383; (2006), p. 8). Plainly, under these circumstances, whichever decision is arrived at is purely a matter of probabilistic chance. But equally plainly, the neural roulette wheel method is not the way real-life deliberators come to a decision. Deliberation is a consciously controlled process that adjudicates between different possible alternative outcomes, each of these being outcomes of the very process that adjudicates between them. (The controlling process controlled: an instance of conscious control-reflexivity). To introduce the analogy of a neural roulette wheel into the analysis of decision making is, I contend, to replace agential autonomy by heteronomy. Roulette wheels cannot coexist with control, and although deliberational processes are essentially indeterministic in the sense of having multiple possible outcomes they are closely shaped and guided by the rational consciousness of their agent-controllers. Decision-making is indeterministic, but its result is not arrived at by chance or luck. Whatever the result, there will be a reason for it, although the weighting of that reason and its comparative weight against other competing reasons may vary from case to case. If God were to order a "replay" of a deliberational process (cf. van Inwagen (2000), reprinted (2002), pp. 172–3), the eventual decision might be different, but the newly-shaped steps that led up to that decision would once more demonstrate the deliberator's autonomy and control.

To sum up, focusing on decisions, arrived at through deliberation, as the results of indeterministic processes, rather than as isolated indeterministic events, blocks their being described as *chance* or *lucky* occurrences. In point of fact, the connectedness of the world enables every event to be regarded in this way, as the culmination of a long or short process, whether deterministic or indeterministic. Deterministic processes are such that their initial conditions plus the laws of nature permit of only one physically possible outcome. Indeterministic processes require two or more. Whether the event that terminates the process is describable as a *chance* event depends on whether or not the process is uncontrolled, or on the other hand fully or partially controlled by a conscious, rational agent.

References

Aguilar, J. and Bukareff, A. eds (2010) *Causing Human Actions: New Perspectives on the Causal Theory of Action*, Cambridge, MA: MIT Press.

Ayer, A.J. (1946) "Freedom and necessity", reprinted in Ayer (1954).

—— (1954) *Philosophical Essays*, New York: St. Martin's Press.

Chisholm, R. M. (1966) "Freedom and action", in *Freedom and Determinism*, ed. K. Lehrer, pp. 11–44. New York: Random House.

Clarke, R. (1993) "Toward a credible agent-causal account of free will", *Nous* 27, pp. 191–203.

—— (1995) "Indeterminism and control", *American Philosophical Quarterly* 32, pp. 125–138.

—— (2003) *Libertarian Accounts of Free Will*, New York: Oxford University Press.

Davidson, D. (1963) "Actions, reasons, and causes", *Journal of Philosophy* 60, pp. 685–700, reprinted in Davidson (1980).

—— (1980) *Essays on Action and Events*, New York: Oxford University Press.

Fischer, J.M. (1994) *The Metaphysics of Free Will*, Cambridge, MA: Blackwell.

—— and M. Ravizza (1998) *Responsibility and Control*, New York: Cambridge University Press.

Ginet, C. (1990) *On Action*, New York: Cambridge University Press.

Goldman, A. (1970) *A Theory of Human Action*, Englewood Cliffs, NJ: Prentice-Hall.

Hobart, R.E. (1934), "Free will as involving determination and inconceivable without it", *Mind* 43, pp. 1–27.

Kane, R. (1999) "Responsibility, luck and chance: Reflections on free will and indeterminism", *Journal of Philosophy* 96, pp. 217–40.

—— (2002) ed. *The Oxford Handbook of Free Will*, New York: Oxford University Press.

—— (2005) *A Contemporary Introduction to Free Will*, New York: Oxford University Press.

McCall, S. (1999) "Deliberation reasons and explanation reasons", in Jackendoff et al. (eds) *Language, Logic, and Concepts*, pp. 97–108. Cambridge, MA: MIT Press.

—— (forthcoming) "Controlled indeterministic processes in action theory", in D. Vanderveken (ed.) *Attitudes and Actions in Discourse*.

McCall, S. and E.J. Lowe (2005) "Indeterminist free will", *Philosophy and Phenomenological Research* 70, pp. 681–90.

Mele, A. (2003) *Motivation and Agency*, New York: Oxford University Press.

—— (2005) "Libertarianism, luck, and control", *Pacific Philosophical Quarterly* 86, pp. 381–407.

—— (2006) *Free Will and Luck*, New York: Oxford University Press.

Nozick, R. (1981) *Philosophical Explanations*, Cambridge, MA: Harvard University Press.

O'Connor, T. (2000) *Persons and Causes*, New York: Oxford University Press.

Pink, T. (2004) *Free Will: A Very Short Introduction*, New York: Oxford University Press.

Searle, J. (2004) *Mind: A Brief Introduction*, New York: Oxford University Press.

Singh, S. (1997) *Fermat's Last Theorem*, London: Fourth Estate.

Smart, J.J.C. (1961) "Free-will, praise and blame", *Mind* 70, pp. 291–306.

Strawson, G. (1986) *Freedom and Belief*, New York: Oxford University Press.

Taylor, R. (1966) *Action and Purpose*, Englewood Cliffs, NJ: Prentice-Hall.

van Inwagen, P. (1983) *An Essay on Free Will*, New York: Oxford University Press.

—— (2000) "Free will remains a mystery", *Philosophical Perspectives* 14, pp. 1–19, reprinted in Kane (2002).

{ 8 }

Controlled Indeterministic Processes in Action Theory

A common criticism of libertarian theories of action is that if what we do is not the result of an unbroken sequence of causes and effects, then it must to some degree be the product of chance. But in what sense can a chance act be intentional or deliberate, in what sense can it be based on reasons, and in what sense can a person be held responsible for it? If free and responsible action is incompatible with determinism, must it not equally well be incompatible with indeterminism? No. This chapter argues that basing action theory on a new category, that of a *controlled indeterministic process*, resolves a variety of classical dilemmas and opens the way to a new understanding of the relationship between actions, reasons, causes, and responsibility.

1. The Nature of CI-Processes

What is a controlled indeterministic process, a "CI-process" for short? I start off by giving examples, and try to explain in what sense they are indeterministic and in what sense they are controlled. A good first example is walking down a crowded sidewalk. This is indeterministic because other walkers come at you unpredictably from different directions, yet it is not like Brownian motion because rarely if ever do you bump into someone. Both you and the other walkers are highly self-controlled. Other examples are skiing, singing, throwing darts, putting on a putting green, driving a car, and playing the piano. Skiing is indeterministic because irregularities in the snow prevent any turn from being completely smooth. In addition, there are marked differences between skiing in control and skiing out of control. The same goes for driving a car. Even for Tiger Woods, the holing of short putts is anything but a science, and throwing darts can become progressively less controlled as the evening progresses. A 100% deterministic piano-playing machine, not pre-programmed but reading the music and depressing the keys mechanically, might be able to give a fair rendition of Bach. But could it deal with the expressiveness demanded by Brahms? Every full-blooded performance of the notorious third Rachmaninov concerto contains unpredictable surprises, while at the same time exhibiting superb control.

An example of a long-lived CI-process would be the composition of a novel. When Tolstoy started *War and Peace*, was it already established when Pierre and Natasha would marry? Given the initial conditions prevailing at the start of any creative composition, including the writing of this chapter on CI-processes, there are always alternative themes, emphases and conclusions which remain possible but unactualized until the last word is written. A process with alternative possible end-points is necessarily indeterministic. Yet a good writer always controls her material, while operating within the framework dictated by the logic of her work. The same goes for a Rembrandt or a Picasso.

Here are some more CI-processes: throwing a baseball, holding one's breath (indeterministic as to the exact length of time until the next breath), tying a bowtie, doing a handstand, swimming, making an omelette, adding up a column of figures[1], taking the square root of 236073, whistling, apologizing. It seems plain that those who execute a CI-process exhibit a skill or technique, whether learned or (in the case of a bird building a nest) innate. But no manifestation of a skill is 100% perfect. Each performance contains idiosyncracies reflecting the indeterminism and the different varieties of control found in all CI-processes, characteristics which set them apart from non-CI-processes such as the algorithmic operation of a Turing machine or of a modern computer.

2. CI-Processes That Result in Decisions

The CI-processes discussed so far, such as walking, whistling or sinking a putt, can all be classified as actions. Even writing a novel can be looked on as an extended multi-year action. But there is another important sub-class of CI-processes which are not themselves actions, but which result in actions, namely deliberative processes which culminate in decisions. Examination of this subclass yields the sought-for linkages between reasons, causes and responsibility that action theory seeks.

When an agent deliberates, the deliberative process she engages in can be divided into three distinct stages:

1. Listing the alternative options
2. Evaluating them
3. Choosing which option to realize

---

[1] "I add them up from top to bottom, then add them up from bottom to top and split the difference". (Stephen Neill)

For example, Ann this evening is faced with the choice of (A) eating out, (B) going to a movie, (C) working on a new edition of prints. Each one of these possibilities is do-able, each excludes the others, and each of (A), (B) and (C) has reasons for, and reasons against, choosing it. In stage 2 of the deliberative process Ann weighs her reasons for, and her reasons against, each option. This process is not like weighing sugar. The reasons do not come with ready-made objective weights, but must be assigned a weight, i.e. weight*ed* by the deliberator. The assigning of weights is controlled by Ann, and may change in the course of the deliberation as new information and new considerations come to mind. For example, there are conflicting reports about the movie; Hugo liked it but Tita did not. Ann must first weight and then weigh the pros and cons of (A), (B) and (C) against one another.

Stage 2 of Ann's deliberation ends with a list of the three options in order of preference: first (B), second (C), third (A). The final stage 3, consisting of the decision to implement (B), may follow immediately, or may be delayed until the last moment for arriving at the cinema on time. In any case Ann acts, when she acts, for a reason, and the causal influence that the reason possesses is determined by the overall control she has over her decisional process.

Since the weights attached to inputs to any deliberative process are controlled by the deliberator, two CI-processes with identical initial conditions and identical inputs may result in different decisions, and consequently different actions. This fact constitutes their indeterministic nature. The indeterminism does not lie in a single undetermined *event*, such as the chance decay of a free neutron or other unstable particle in quantum mechanics, or even in a series of such events. Instead, deliberation is an indeterministic *process*, not an event. Its indeterminism is continuous rather than discrete, and derives from the continuous nature of the control exercised by the deliberator in weighting the various options, and the importance he or she attaches to other inputs during the course of the deliberation. In a deliberative CI-process, reasons are causes, but they are not sufficient causes. They are causes that in Leibniz's phrase "incline without necessitating". That they do not necessitate is due to the overall control exerted by the deliberator.

Because reasons incline but do not necessitate, it is possible to give an "event-causal" rather than an "agent-causal" account of deliberation and action, where the causes are reasons and the things of which they are causes are actions.[2] Since reasons are not sufficient causes, the variety of causation appropriate to deliberation and decision is *probabilistic causation*. One could

[2] See Randolph Clarke, *Libertarian Accounts of Free Will* (2003), and Timothy O'Connor, *Persons and Causes* (2000). Alfred R. Mele, in *Free Will and Luck* (2006) p. 10, remarks that "Typical event-causal libertarianism encompasses a commitment to what may be termed *agent-internal indeterminism*." But Mele himself stops short of embracing a full-blown indeterministic event-causal account of action based on reasons as probabilistic causes.

view the causality of reasons in deliberation as a probability function, in which the probability of each option varies continuously and the probabilities p(A), p(B), p(C),... sum to one. When the moment of decision arrives, one of these probabilities takes the value 1 and the others become 0. Throughout deliberation the different probability values vary smoothly, in a reciprocal way.

As was stated in section 1 above, viewing deliberation as a controlled indeterministic process allows the concept of reasons-based action to occupy an intermediate position between chance and determinism. A reasons-based action is not the product of a 100% deterministic process, nor is it the outcome of a single random event or series of random events. Instead, it is the end-product of a process that is both continuous and indeterministic. The indeterminism does not derive from the process's containing isolated chance events, but from its having different physically possible terminations. Reasons-based actions consequently pass between the Scylla of determinism and the Charybdis of chance.

3. Control

"Control" is a slippery word, and we need to be precise about its different meanings.[3] Take the example of steering a car. We would say that the driver *controls the car's steering* if a necessary and sufficient condition of its turning to the right is that the driver turns the wheel to the right, and a necessary and sufficient condition of the car's turning to the left is that the driver turns the wheel to the left. Having said this, it is plain there exist lesser degrees of control. Suppose that the steering mechanism is defective in such a way that when the wheel is turned right, the car goes right, but there may be instances in which the car turns right *without* the wheel being turned right. That is, turning the wheel right is only a *sufficient* condition of the car's turning right, not a *necessary* condition. I shall say that in this case the driver exercises "performance control" over turning right. Alternatively, turning the wheel right may be only a necessary condition of the car's turning right, not a sufficient one. That is, sometimes the car fails to turn right when the wheel turns right, although it never turns right *unless* the wheel turns right. In this case I say that the driver exercises "omission control" over turning right, so-called because omitting to turn the wheel prevents the car's turning. "Full control" requires both performance and omission control.

---

[3] See J.M. Fischer, *The Metaphysics of Free Will* (1994), pp. 132–34, and J.M. Fischer and M. Ravizza, *Responsibility and Control* (1998), pp. 30–32. The account given here differs significantly from Fischer's. In Aristotle's *Ethics* 1102b14, self-control is "the disposition to refuse to act on desires not approved by reason; the lack of it...is the disposition to give way to them". (Broadie and Rowe, *Aristotle: Nicomachean Ethics* (2002), p. 295.) See also Mele, *Autonomous Agents* (1995), p. 4 ff., and his *Irrationality* (1987).

Affinities exist between performance control and what Fischer calls "guidance control", and between full control and what he calls "regulative control". But much still remains to be said about these different levels of control. In the field of human psychopathology, one would not like to be in the position of exercising only performance control, but not omission control, over the movement of one's limbs. However, since thoughts frequently enter one's head unbidden, one may be precisely in that position concerning the operations of the mind. We can think what we want to think, but sometimes we think what we don't want to think. Despite all this, the variety of control we possess over deliberation, including the weighting of reasons and the power of whether, when, and what to decide, is "full" control—that is, the combination of both performance and omission control.

4. Responsibility

Under what conditions can someone who does A be said to be responsible for doing A? This is a huge subject, discussed in dozens of papers in recent years, most of which owe their inspiration to Harry Frankfurt's seminal 1969 paper "Alternate possibilities and moral responsibility". Frankfurt's "Principle of alternate possibilities" (PAP) states that a person is morally responsible for what he has done only if he could have done otherwise. Frankfurt argues persuasively that PAP is false, and that someone may well be responsible for what he has done even though he could not have done otherwise. From here it is only a short step to the conclusion that moral responsibility is compatible with physical determinism.

Frankfurt's counter-example to PAP is well-known. Black wants Jones to perform a certain action A that Jones is contemplating performing in any case, and arranges things so that if Jones is tending towards a decision *not* to do A, Black will create in him an irresistible inner compulsion to do A all the same. (For instance, Black may hypnotize Jones, or may manipulate his neural processes using electrodes implanted in his brain.) Jones is unaware of these arrangements, and in the end decides to do A without Black's fail-safe device coming into play. In such a situation, Jones is responsible for doing A, even though he could not have done otherwise. Consequently PAP is false.

Many philosophers have disagreed with Frankfurt's reasoning, some arguing that at the time Jones decided to do A he could have decided otherwise, and that our judgement concerning his responsibility is based on this initial ability to do, or not to do, A. Only later, when Black's device came into play, would Jones have had no choice.[4] In support of Frankfurt, others have proposed new

---

[4] See Robert Kane, *The Significance of Free Will* (1996). Kane notes that Frankfurt's whole example is presented within an indeterministic framework, neither Black, nor Jones before the

and ingenious "Frankfurt-style examples", in some of which Jones' decision to do A and Black's fail-safe device are stipulated (perhaps unrealistically) to become operative at exactly the same instant.[5] At that point it becomes difficult to say which prevails; whether Jones' decision pre-empts Black's device, or whether the device pre-empts Jones' decision. My purpose here is not to pursue this point, but rather to suggest a different strategy for libertarians, a strategy that is immune to Frankfurt-style counter-examples.

Frankfurt's indeterministic principle PAP states a necessary condition for moral responsibility. A person is morally responsible for what he has done, only if he could not have done otherwise. But if we add to indeterminism the notion of control, then it becomes possible to give not only a *necessary* condition for responsibility, but a *necessary and sufficient* condition. This strengthened condition resists counter-exampling in the style of Frankfurt. Or so I shall argue.

5. Necessary and Sufficient Conditions for Responsibility

To replace PAP I suggest the following "Principle of Responsibility and Control" (PRC). The principle concerns responsibility in general, not the more restricted concept of moral responsibility.

> PRC. *A person who performs action A at time t is responsible for having done A if and only if, throughout some time interval immediately before t, the doing of A, including whether or not to do A, was in her control.*

Although it is not stated explicitly, it will be clear that PRC applies only to actions that originate within an indeterministic context. That is, it applies to actions that result from a CI-process, where the controller of the process is the agent. If the question of whether or not to do A falls within the agent's control, then ipso facto the deliberative process has at least two different possible outcomes, doing A and not doing A, and is therefore indeterministic.

Is PRC open to Frankfurt-style counterexamples? It would seem not. A counter-example would involve a controlled action of an agent *x* for which *x* is not responsible, and it seems implausible that there are any such. Alternatively, it would involve a performed action, not within *x*'s control, for which *x* is

decision, being causally necessitated to do what he does. Can an indeterministic example be used to argue against indeterminism?

[5] See Alfred R. Mele and David Robb, "Bbs, magnets and seesaws: The metaphysics of Frankfurt-style cases", in *Moral Responsibility and Alternative Possibilities* (2003), edited by David Widerker and Michael McKenna, a volume of papers on Frankfurt and PAP. See also Carl Ginet, "On Mele and Robb's indeterministic Frankfurt-style examples", in *Philosophy and Phenomenological Research* 80 (2010), pp. 440–46; Michael McKenna, "Frankfurt's argument against alternative possibilities: Looking beyond the examples", in *NOUS* 42 (2008), pp. 770–93; and David Widerker, "Libertarianism and the philosophical significance of Frankfurt scenarios", *Journal of Philosophy* 103 (2006), pp. 163–87.

nonetheless responsible, and this seems equally implausible. However, it is not out of the question that some Frankfurt of the future may one day uncover an ingenious counter-example to PRC, and I leave the possibility open as a challenge. The remainder of this section is devoted to examining cases which are *not* counter-examples to PRC, but which support the principle.

Imagine the following case, adapted from J.L. Austin.[6] I go to shoot my donkey, but just as I pull the trigger you jog my arm, with the result that I shoot *your* donkey at time *t*. Am I responsible for shooting it? No, because during a short time interval before *t* the shooting of it was not under my control. It was an accident. This case differs from another Austin example in which I go to shoot my donkey, but because of the similarity of the animals I shoot yours instead, not by accident but by mistake. In this case, I *am* responsible for shooting and killing it, but as I argue in the next paragraph I may or may not be *morally* responsible for its death.

As was mentioned at the beginning of this section, PRC provides necessary and sufficient conditions not for *moral* responsibility, but for responsibility *simpliciter*. If we are dealing with the concept of responsibility this is surely the way to go: first explicate responsibility, then deal with morality. The trouble is, that in some cases the two categories of "responsibility for doing A", and "moral responsibility for doing A", are so intertwined that it is difficult if not impossible to separate them. Let us try however.

Consider Elizabeth Anscombe's example of moving the pump handle up and down, and in so doing poisoning a house's water supply.[7] I may be responsible for performing action A, but am I responsible for performing A under *every* description of A? My (Davidsonian) answer is yes. If the action of moving the pump handle up and down is the same action as poisoning the house's water supply, then I am responsible for poisoning the water supply. This holds even though I may be unaware that I am performing that reprehensible deed—that is, that the reprehensible description applies to it. On the other hand, I may not be *morally* responsible for poisoning the water supply. Examples like this demonstrate the importance of distinguishing between responsibility and moral responsibility.

Another more difficult example, along similar lines, is the following, from H.L.A. Hart and A.M. Honore's *Causation in the Law* (1959), p. 72. If *x* puts poison in *y*'s soup and *y* drinks it and dies, then *x* is responsible for *y*'s death. But suppose *y knows* that the poison is in the soup and drinks it anyway? Does *y*'s free, intervening and knowing action break the causal chain that links *x*'s action to *y*'s death? I confess I don't know in this case what to say. Maybe *x* is responsible for *y*'s death, but not morally responsible. Or maybe *x* is not even responsible *simpliciter*, let alone morally responsible. "To consequences no limit

[6] J.L. Austin, "A plea for excuses", in his *Philosophical Papers* (1961), p. 133.
[7] G.E.M. Anscombe, *Intention* (1963), p. 37.

can be set". The causal consequences of our actions are endless, and causal chains can extend for hundred of years into the future. Are we responsible for all the consequences of what we do? Surely not. Yet where should we draw the line? I don't know. I do however believe that the concept of responsibility *simpliciter* should be kept separate from that of moral responsibility, and legal responsibility, and "capacity responsibility", and from other notions such as Hart's concept of "role responsibility". Let me conclude with the following marvellous passage from Hart's *Punishment and Responsibility* (1968), p. 211, illustrating the diversity of responsibility-concepts. I would hope that PRC might serve to extract one single thread of what may be called "action responsibility" from the complex tangle found in Hart's passage. But I am not optimistic. There are still too many additional categories of responsibility,for instance responsibility for omissions, negligence, etc., to be confident that our principle PRC covers them all.

"As a captain of the ship, X was responsible for the safety of his passengers and crew. But on his last voyage he got drunk every night and was responsible for the loss of the ship with all aboard. It was rumoured that he was insane, but the doctors considered that he was responsible for his actions. Throughout the voyage he behaved quite irresponsibly, and various incidents in his career showed that he was not a responsible person. He always maintained that the exceptional winter storms were responsible for the loss of the ship, but in the legal proceedings brought against him he was found criminally responsible for his negligent conduct, and in separate civil proceedings he was held legally responsible for the loss of life and property. He is still alive and he is morally responsible for the deaths of many women and children."

6. Rationality

A further topic needs to be addressed: the question of rationality. The process of deliberating and deciding can be done either rationally or irrationally, and consequently there are such things as rational controlled indeterministic processes, or "RCI-processes". What distinguishes a rational from an irrational deliberative process, a rational from an irrational practical decision? At this early stage in our understanding of CI-processes, the question is best addressed by considering concrete examples.

Chess-playing is a quintessentially rational process, which is intellectually demanding, highly controlled, and indeterministic because the opponent's moves are unpredictable. (A "deterministic" chess game, for instance against a computer, in which the computer's moves are calculable and predictable, would not be a game at all.) Chess-playing is an exercise in rational decision-making under uncertainty: an RCI-process.

A very different case of rational decision-making is that of Mele's Beth (Mele, *Free Will and Luck* (2006), pp. 106–7). Beth is a lawyer in a large competitive urban law firm who deliberates about quitting her job and working in a small town where her moral principles would be less likely to be compromised. She visits the small-town firm, is impressed by the enhanced quality of life of the new environment, and is just about to accept a position when it occurs to her it would be prudent to return once more and explore the town on her own. On her second visit she encounters a degree of conservatism and bigotry she was previously unaware of, and she promptly refuses the offer and remains in the city. Beth's decision-making process is a rational one, based on practical rather than intellectual reason, and features a radical change in her weighting of the pros and cons of moving vs. staying after her second visit. The change in weighting is brought by Beth, and exemplifies the rational control she exercises over her deliberation. The latter stands as a typical case of a rational, controlled, indeterministic process.

7. Conclusion

What is new about this chapter is the concept of a controlled but indeterministic process, exemplified by activities like walking, running, writing a poem, playing tennis, etc., but more importantly for action theory by the process of deliberating and deciding. When we decide and act for a reason, that reason, in competition with other reasons on which we do *not* act, serves as a probabilistic cause of the eventual action. If our action is a rational one, the various competing probabilistic causes are controlled by Reason with a capital R. If our action is less than fully rational (as thankfully many actions are), it will in most cases still be controlled by the agent, though the latter may be influenced by desire, emotion, laziness, passion, or fear. Sometimes a Iago can manipulate an Othello, playing upon his weaknesses and sharing the responsibility for whatever crimes Othello commits. In such cases, Othello's control over his own behaviour is diminished, and according to the principle PRC, responsibility should be reduced proportionately to lessening of control. CI-processes provide a framework within which an adequate libertarian account of autonomous, reasons-based, responsible action can be formulated.

References

Anscombe, G.E.M. (1963) *Intention*, Ithaca, NY: Cornell University Press.

Austin, J.L. (1961) *Philosophical Papers*, New York: Oxford University Press.

Broadie, S. and Rowe, C.J. (2002) *Aristotle: Nicomachean Ethics*, New York: Oxford University Press.

Clarke, R. (2003) *Libertarian Accounts of Free Will*, New York: Oxford University Press.

Fischer, J.M. (1994) *The Metaphysics of Free Will*, Cambridge, MA: Blackwell.

—— and Ravizza, M. (1998) *Responsibility and Control*, New York: Cambridge University Press.

Frankfurt, H. (1969) "Alternate possibilities and moral responsibility", *Journal of Philosophy* 66, pp. 829–839.

Ginet, C. (2010) "On Mele and Robb's indeterministic Frankfurt-style examples", *Philosophy and Phenomenological Research* 80, pp. 440–46.

Hart, H.L.A. (1968) *Punishment and Responsibility*, New York: Oxford University Press.

—— and Honore, A.M. (1959) *Causation in the Law*, Oxford: Clarendon Press.

Kane, R. (1996) *The Significance of Free Will*, New York: Oxford University Press.

McKenna, M. (2008) "Frankfurt's argument against alternative possibilities: Looking beyond the examples", *NOUS* 42, pp. 770–93.

Mele, A.R. (2006), *Free Will and Luck*, New York: Oxford University Press.

—— and Robb, D. (2003) "Bbs, magnets and seesaws: The metaphysics of Frankfurt-style cases", in D. Widerker and M. McKenna (ed.) *Moral Responsibility and Alternative Possibilities*, Burlington, VT: Ashgate.

O'Connor, T. (2000) *Persons and Causes*, New York: Oxford University Press.

Widerker, D. (2006) "Libertarianism and the philosophical significance of Frankfurt scenarios", *Journal of Philosophy* 103, pp. 163–187.

{ 9 }

The Causative Power of Conscious Choice

1. Causal Closure and Causal Exclusion

The main obstacles standing in the way of giving an account of mental causation are two plausible and wide-ranging principles applying to the physical world: causal closure and causal exclusion. PCC, the principle of causal closure, states the following:

PCC: If a physical event has a cause at *t*, it has a physical cause at *t*.

Causal exclusion is captured by PCE:

PCE: If an event *e* has a sufficient cause *c* at *t*, no event at *t* distinct from *c* can be a cause of *e* (unless this is a genuine case of causal overdetermination).

Jaegwon Kim, who has arguably wrestled more than anyone else since Descartes with the problem of mental causation, formulates these two principles on pp. 15 and 17 of his (2005). Taken together, they effectively rule out any possibility of mental causation. If a physical event such as the movement of my hand in writing the word "pen" has any cause at all, then by PCC it has a physical cause involving motor neurons, the muscles of my arm, etc. But if these bodily causes are sufficient for the production of the word "pen", then, by PCE, postulating an additional mental cause, such as the intention to write "pen", results in causal overdetermination. The physical world is causally closed, and so-called mental causes contribute nothing to it. Mental causes are physically inert. The philosophical problem of mental causation is to see how, if at all, this difficulty may be overcome.

2. Indeterministic Neural Events

Since by PCC physical events, if they have causes at all, have physical causes, there could exist a non-redundant mental cause of a neurophysiological event

X only if X had no physical cause, i.e. were indeterministic. Do indeterministic neurophysiological events exist? Yes. It can be shown that the process of practical deliberation, i.e. deliberation about what to do, implies the existence of such events. Practical deliberation is the process whereby an agent, faced with two or more alternative courses of action, selects one of them as her preferred choice. The entire process, which can last anywhere from one or two seconds to days, weeks, or even months for important decisions that require careful thought, falls into three separate temporal phases, as follows (McCall (1987), (2008); McCall and Lowe (2005)). As will be seen the overall process is an indeterministic one.

(i) *Choice set formation.* The first stage of any deliberation is to list the alternatives being considered, the "choice set", containing a minimum of two members. Coffee or tea? Milk or sugar, or both? More serious deliberations may involve whether (A) to study microbiology, (B) to spend a year as a Red Cross volunteer, or (C) to get a job.

(ii) *Evaluation.* The pros and cons of each alternative are weighed against one another, and a list of the items in the choice set is constructed, in order of preference.

(iii) *Decision.* Given the preferential ordering of the alternatives, there remains a final step in any deliberation, namely decision or choice. This may or may not follow the preferential list: one can succumb to Aristotelian weakness of will and "know the good but not do it", although normally the option that heads the list will be the chosen one.

An essential characteristic of the process of deliberation is that given the conditions prevailing at its start, or at any time before the final choice, two or more different outcomes are always available and possible. Each alternative course of action that is a member of the choice set remains "open", i.e. choosable or do-able, throughout the entire process. Let A, B and C be deliberated-upon alternatives, each one initiated by a different bodily movement, and let n(A), n(B) and n(C) be the motor neuronal discharges producing movements of the body that lead to A, B and C respectively. For example, neuronal discharge n(A) might lead to the signing of an application form for admission to a graduate degree program in microbiology, n(B) to a volunteer letter for the Red Cross, n(C) to a phone call applying for a job, etc. As long as the deliberative process continues, each of the physical neuronal events n(A), n(B) and n(C) remains "potential", i.e. physically possible but not actual. Furthermore they are mutually incompatible; if one takes place its occurrence excludes the other two. The end of deliberation lies in the actualization of one of the original alternatives.

What is asserted in the preceding paragraph can be summarized by saying that deliberation is an indeterministic process. Any process with two or more physically possible, mutually incompatible outcomes is indeterministic. What resolves the indeterminism is the eventual decision or choice that terminates deliberation. This event, leading to one of the motor neuronal discharges n(A), n(B), n(C) in the example considered, is an event without a physical cause. It is in fact a *conscious* choice, and serves as the mental cause of whichever one of the physical events n(A), n(B), n(C) gets realized. Since none of n(A), n(B), n(C) is physically caused, the principle of causal exclusion does not come into play, and there is no causal overdetermination. Conscious choices or decisions that terminate deliberations are not physically inert, but have concrete physical effects. They affect what happens in the world. Mental causation is a reality that we all live with, and its possibility rests on the indeterministic character of Aristotle's *bouleusis*, practical deliberation. Not even a Laplacian demon could predict, at the beginning of a deliberative process, what the outcome would be. The conscious decision that brings it to an end could perhaps be predicted by a psychiatrist, or by a friend who knows the deliberator well, but not by a demon whose predictive expertise is confined to the world of physics. The physical realm is not causally closed, but from time to time is open to mental causes.

3. Mind-Body Supervenience

Given that on occasion conscious, mental events can cause neurophysiological events and movements of the body, it may be asked, what should be made of elaborate philosophical theories designed to show that though the mind may exist, it is physically powerless? Chief among these is the theory of mind-body supervenience, according to which mental properties supervene on physical properties and are dependent upon them. Thus Kim:

> Mental properties *supervene* on physical properties, in that necessarily, for any mental property *M*, if anything has *M* at time *t*, there exists a physical base (or subvenient) property *P* such that it has *P* at *t*, and necessarily anything that has *P* at a time has *M* at that time. ((2000), p. 9).

Once supervenience is accepted, mental causation in the full-blooded sense of a mental event M directly causing a physical event P becomes impossible. At best, M could cause P indirectly in virtue of supervening upon some other physical event P*, which in turn causes P. But this is not mental causation, as exemplified in the preceding section by a conscious mental choice causing the activation of one of the three motor neuronal events n(A), n(B), n(C). The conscious choice is not itself a physical event at all, and none of n(A), n(B), n(C) is physically caused. If one of them were, its occurrence would be in principle predictable, and the deliberative process would not be indeterministic. Unpredictability, indeterminism and mental causation hang together in this argument, and mental causation would be impossible in a 100% deterministic world because of physical causal closure and exclusion.

Returning to mind-body supervenience, there seems no alternative but to reject it if mental causation is real. If mind and consciousness have sufficient robustness and ontological independence to penetrate the causal closure of the physical realm, they cannot merely supervene upon physical events and properties of the central nervous system. In a way this is a pity, because supervenience is an elegant and attractive theory of the mind. But it conflicts head-on with the causal power of conscious choice as manifested in the process of deliberation.

4. What Variety of Cause Is a Conscious Choice?

It remains to consider what kind of causation is in question when a conscious choice causes one of the physical events n(A), n(B), n(C) to take place, and a neural impulse activates one of the muscles of the body. The verb "choose" in this particular context is an achievement verb, as distinct from a task verb (Ryle (1949), pp. 150–53), and the perfect passive participle "chosen" implies that the achievement has been accomplished and the relevant option successfully and irrevocably selected. In the same fashion, the perfect passive participle "seen" implies the success of the achievement verb "see". (The task verb "look", on the other hand, lacks a passive participle entirely.) Once something has been seen, or once an option has been chosen, the game is over. To revoke a choice is an elaborate procedure, one that in the context of practical deliberation is impossible since the relevant motor neuron has already fired. What's done cannot be undone. It would be easier to revoke one's vote in a national election than to change an option from "chosen" to "unchosen" at the end of a deliberation. Conscious choice, in this context, exercises an irrevocable causative power.

Still, questions remain. Mental causation of the conscious choice variety would appear to be anomalous, not based on any causal law we are familiar with. In a trivial sense it supports counterfactuals, but only of the uninformative type "If option A had been consciously chosen, then motor neuron n(A) would have fired". Perhaps the best way of characterizing conscious choice causation would be to describe it as "Humean". What does this mean? Hume's well-known analysis of causation is divided between the *Treatise* and in the *Enquiry.* In the former he says (i) that "whatever are consider'd as causes or effects, are *contiguous*", and cannot be "ever so little removed" in either time or place from one another. His second requirement (ii) is "Priority of time in the cause before the effect." (*Treatise* I, III, sect. 2, Selby-Bigge pp. 75–76.) The third and most famous criterion is (iii) constant conjunction: "we may define a cause to be *an object followed by another, and where all the objects, similar to the first, are followed by objects similar to the second*". (*Enquiry* sect. VII.) This is the well-known Humean position, but the author unexpectedly follows it up with: "Or, in other words, *where, if the first object had not been, the second never had existed.*" The latter is certainly not a case of "in other words", and Hume's follow-up is far from a re-statement of criterion (iii), constant conjunction. If the presence of X is always followed by the presence of Y, it is plausible to say that X is a "sufficient condition" of Y. But (iv) if the *absence* of X is invariably followed by the *absence* of Y, then X is a "necessary condition" of Y. Together, (iii) and (iv) imply X's being a necessary and sufficient condition for Y, and perhaps this is exactly what Hume had in mind.

Applying these Humean ideas to the relationship between conscious choice and motor neuronal activation, criterion (i), contiguity in time and place, has to be modified to "contiguity in time" only, since conscious choice has no spatial location (*pace* mind-body supervenience). Conscious thoughts are located in time but not in space. Criterion (ii), priority in time, is relevant: mental causes must precede their physical effects. Criteria (iii) and (iv) seem also to be entirely applicable. It is fitting to regard a conscious choice X as both a necessary and a sufficient condition for the neuronal discharge Y of a given motor neuron, and for the subsequent contraction of a muscle leading to a bodily movement. The upshot is, that it is not inappropriate to describe as "Humean causation" the relationship between conscious choice and action occurring at the end of a deliberative process.

References

Kim, J. (2000), *Mind in a Physical World*, Cambridge, MA: MIT Press.

—— (2005), *Physicalism, or Something Near Enough*, Princeton: Princeton University Press.

McCall, S. (1987), "Decision", *Canadian Journal of Philosophy* 17, pp. 261–288.

—— (2008), "How to make a decision", in *Actions, Rationality and Decision*, ed. Vanderveken and Fisette, pp. 325–338. London: College Publications. See Chapter 4 above.

McCall, S. and Lowe, E.J. (2005), "Indeterminist free will", *Philosophy and Phenomenological Rescrarch* 70, pp. 681 690.

Ryle, G. (1949), *The Concept of Mind*, New York: Hutchinson's University Library.

{ 10 }

The Open Future and Its Exploitation by Rational Agents

Branching along the time dimension provides a dynamic, four dimensional, tree-like spacetime structure that explains many features of the physical world. The list includes temporal asymmetry and directionality, time flow and the existence of "now", physical versus logical possibility, the openness of the future, quantum probabilities, and superpositional collapse. The chapter discusses these, and how agents use the open future in order to act intelligently and rationally.

1. Branching along the Time Axis

A single four-dimensional manifold or "history" of the world (a "Minkowski world", or, to allow for general relativity, a curved spacetime manifold) extends from the Big Bang to the end of time if there is one, or indefinitely, if there is not. A "slice" of a history is a three-dimensional instantaneous state of such a manifold. In branching spacetime, histories branch along such instantaneous slices, i.e. "spacelike hypersurfaces", and the branching is towards the future, not the past. (The fact that there is no branching towards the past reflects our conviction that the world has a *unique* past. It could not be true both that the man from Stratford was the author of *Hamlet*, and also that Bacon was.) A fan of branches above the first branch surface connects with a single history extending below. The overall structure is that of a tree, with an unbranched trunk up to a certain point and a multitude of branches above that. Since the branches themselves branch upwards, the branching is very dense.[1]

[1] McCall (1994) contains a detailed account of the branching spacetime model discussed here. Belnap (1992), and Belnap et al. (2001), introduce a similar model, the details of which differ from McCall's. The two models bear only a superficial resemblance to the Everett-Wheeler many-worlds interpretation of quantum mechanics.

2. The Direction and Flow of Time

Because branching is only towards the future, temporal asymmetry and directionality are built into the structure of the universe. By definition, the "past" is constituted by the single trunk, the "future" by the branches, and the "present" by the lowest branch point or branching hypersurface on the tree. The "flow" of time consists of the progressive movement of the first or lowest branch point up the tree, brought about by "branch attrition".[2] Branch attrition consists in the following. Of all the branches which split off at the lowest branch point, one and only one is selected to become part of the trunk, and the others vanish.

The progressive disappearance or vanishing of all branches but one, at the lowest branch point of the tree, results in a new branch point becoming lowest and therefore "present". This is a fairly heavy-laden metaphysical idea. It is both ontological and dynamic, and is not intended as a metaphorical or analogical description of the world, but as literal and precise. In the branch model presented here, the branches really do fall off in the way described, and their falling off constitutes the flow of time. Their progressive disappearance can also be compared to Aristotle's "transition from potentiality to actuality", which in modern physics consists of the collapse of the wave-function brought about by interaction with a measuring device (Heisenberg (1958), pp. 53–58). Unlike the "mind dependence" theory of temporal becoming (see Grünbaum (1963), chapter 10), in which the passage of time is a subjective illusion, time flows in the dynamic branched universe whether conscious beings perceive it or not. This is illustrated by figures 1 and 2 below:

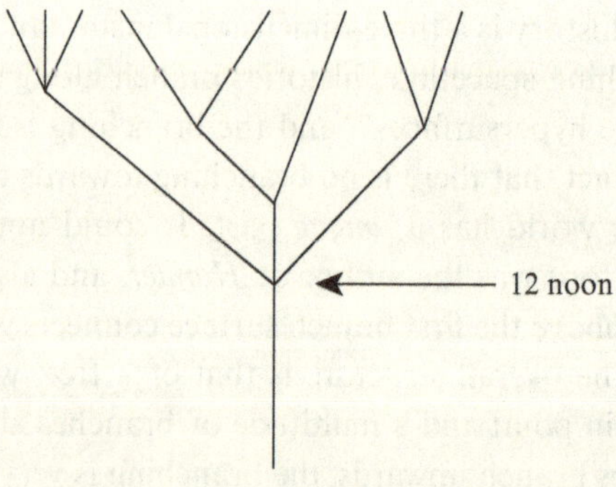

FIGURE 1

[2] For an account of how branch attrition in the model corresponds to the flow of time, see McCall (1976), (1984), (1994), and (1997).

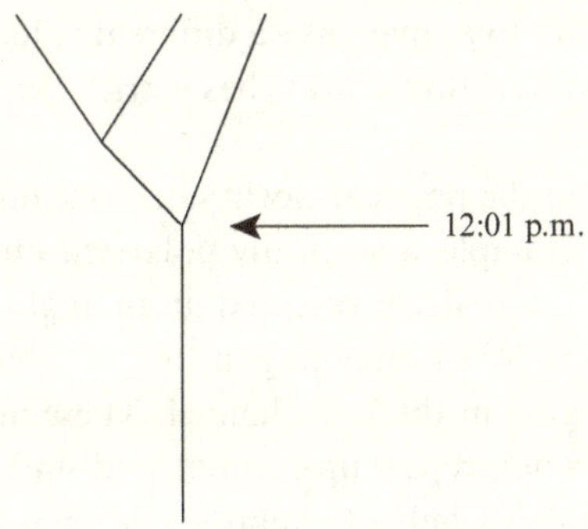

FIGURE 2

3. Physical Possibility and Open-ness

Indeterminism is built into the model. In the set of all branches that extend into the future at the first branch point, each has an equal chance of being selected as the "actual" branch that becomes part of the past. At the level of the present, branch selection is random. Branching in the model is discrete rather than continuous, there being a small finite, non-zero interval between successive branch points. A future event is *physically possible*, relative to the present state of the world, if it occurs on some future branch. More precisely, E at time t_2 is possible, relative to conditions prevailing at a branch point or branch surface X at time t_1, if E at t_2 is on some branch above X at t_1. For example, it is physically possible, in 2010, for a traveler in Montreal to be in Vancouver five hours later, but it is not possible to be there five minutes later.[3] This holds even though it is *logically possible* to be there five minutes later, or five nanoseconds later. The branched model yields a clear, unambiguous difference between logical and physical possibility. The same holds of physical necessity. What is physically necessary, relative to time *t*, is what is on all branches above *t*. It is physically necessary for water heated in an open container above 150°C to boil, but it is not logically necessary. An open future is a branching future that contains mutually incompatible, physically possible events.

4. Probability, and Superpositional Collapse

The *probability* of a future event is given by the *proportionality* of future branches on which the event occurs. For example if there are 100 future branches above the first branch point of a tree at time t_1, and if an event of type E occurs at time t_2 on 67 of them, then p(E-at-t_2), relative to t_1, is 0.67. Of

[3] Concerning the concepts of physical possibility and physical necessity see McCall (1969).

course p(E-at-t_3), relative to t_1, may take a different value. The notion of "proportionality" among sets of future branches is tricky, and needs to be defined precisely.

Some probabilities in the physical world take irrational values, and require careful handling. For example, a vertically polarized photon about to enter a two-channel polarization analyzer oriented at an angle of 34° to the vertical has a probability of $\cos^2 34°$ of emerging in the "+" channel, and a probability of $\sin^2 34°$ of emerging in the "−" channel. These are irrational numbers. $\cos^2 34° = 0.68729\ldots$, a non-repeating decimal, and $\sin^2 34° = 0.31270\ldots$. How can an irrational probability value be represented by a relative proportion of branches above a branch point?

The answer lies in considering a decenary tree, a spacetime tree that splits into 10 branches at the first branch point, with every branch dividing in 10 at each successive branch level.[4] A decenary tree can be compressed into an arbitrarily short but nonzero temporal interval, say Δt, as follows. The tree branches in 10 at $t = 0$, and each branch divides in 10 at time $t + \frac{1}{2}\Delta t$, at $t + \frac{3}{4}\Delta t$, at $t + \frac{7}{8}\Delta t$, . . . etc. By $t + \Delta t$, the decenary tree will contain a non-denumerable infinity of branches. See figure 3 below:

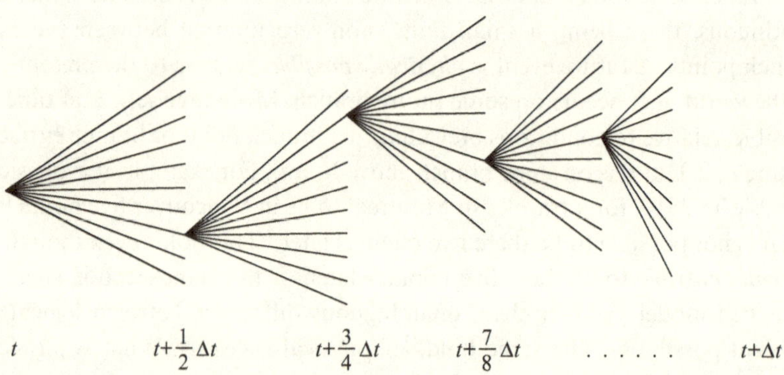

FIGURE 3

How does it come about that exactly $\cos^2 34° = 0.68729\ldots$ of these are + branches, and $\sin^2 34° = 0.31270\ldots$ are − branches? Well, suppose that six of the first ten branches are + branches, three are − branches, and one branch is "open", meaning that it is neither + nor −. Once a branch is +, or −, it stays that way till the top of the decenary tree. At the second level, at $t + \frac{1}{2}\Delta t$, the one open branch divides into eight + branches, one − branch, and one open branch. At $t + \frac{3}{4}\Delta t$, the open branch divides into seven + branches, two − branches, and one open branch. At $t + \frac{7}{8}\Delta t$, the open branch splits into two + branches, seven − branches, and an open branch. And so forth. The

[4] Decenary trees are described in McCall (1994), pp. 88–92.

decimals 0.68729... and 0.31270... are simply reproduced in the decenary tree. At the end, $t + \Delta t$, exactly $\cos^2 34°$ of the totality of branches in the tree will be branches in which the photon exits in the + channel, and $1 - \cos^2 34° = \sin^2 34°$ of the totality will be − branches. In addition there will remain one open branch, which can become arbitrarily either + or − at the initial node of the next level of decenary trees. (When a non-denumerable set is divided into two proper subsets, the addition of one more unit to either of the subsets makes no difference to the overall relative proportionality.)

We may call a decenary tree a "prism", of temporal height Δt. The entire branched spacetime universe is a *prism stack*, with a new prism standing at the upper (i.e. later) end of each complete path through the prism immediately below it. Despite what Georg Cantor says about there being no well-defined relative proportionalities in infinite sets, decenary trees provide exact probability values for future events, even when these values are irrational numbers. The reason why decenary trees are exceptions to Cantor's generalization is because of the particular structure of the set of their branches. Suppose a vertically polarized photon enters an analyzer oriented at 34° to the vertical at time t. The probability of its emerging in the + channel, a result which corresponds to a + branch being the sole survivor of branch attrition in the appropriate prism stack, is precisely 0.68729..., and the probability of its emerging in the − channel is precisely 0.31270....

In quantum mechanics, the incoming photon, about to enter a measuring device (in this case a polarization analyzer), is said to be in a *superposition* of polarized states, written $| + \rangle + | - \rangle$. On measurement, the superposition *collapses* into one of these states. In dynamic branching spacetime, the selection of one single branch out of a multitude of branches at a branch point provides an "objective" theory of superpositional collapse, i.e. a theory of collapse that makes no reference to the existence of an observer.[5]

5. Branching Spacetime and Human Deliberation

As was said earlier, a branching future that contains physically possible, mutually incompatible events is "open". When human beings deliberate over what to do (practical deliberation), or what to believe (cognitive deliberation), there are different options open to them. Let's focus on practical deliberation. I have the choice of leaving early and walking home tonight, or leaving later and taking the Metro. Each alternative is physically possible, and each has its advantages and disadvantages. In branching spacetime, relative to "now", there are walking-branches and Metro-branches, in relative proportions that yield

---

[5] See McCall (2000a) and (2000b).

p(walking) = p_1, and p(Metro) = p_2. Is it a truly random matter as to which "actual" branch is selected, among the huge number of future branches that confront me? If p_2 is considerably larger than p_1, the chances of my taking the Metro would seem to be a lot greater than walking. But is this all that can be said? Do I simply wait to see what set of future branches, the "walk" set or the "Metro" set, the actual branch falls into, much as I wait to see, in the polarization experiment, whether the actual branch turns out to be a "+" branch or a "−" branch? No. In day-to-day living, human beings are agents, not observers. The reason I eventually find myself on a Metro branch, rather than a walking branch, is not that random branch selection resulted in the actual branch falling into the larger of the two sets of future options. Instead, after due deliberation, I *chose* or *decided* not to walk. If the "walk" set had contained only a single branch, could I not have chosen it? Probably I wouldn't, but *couldn't* I have?

The process of deliberation, which Aristotle calls *bouleusis*, consists of three stages:

(i) Listing the alternatives that are open;
(ii) Evaluation, i.e. assigning each alternative a weight, either positive or negative, and then weighing one alternative against another;
(iii) Choosing one alternative, resulting in a bodily movement that realizes it. Aristotle calls this *prohairesis*, deliberative choice.[6]

The history of philosophy has witnessed centuries of debate, beginning with the Stoics and St Augustine, over whether in a world that is deterministic, or in which God knows what the future is going to be, humans can exercise free will. Concerning determinism vs indeterminism, either (i) the action A that we perform is deterministically caused by our beliefs and our desires, which would make the performance of a different action B physically impossible, or (ii) the performance of A is a purely chance event. Both alternatives do violence to our deeply felt conviction that what we do is under our control, in some sense "up to us". Branching spacetime, considered as a preferable replacement for rigid 100% determinism,[7] would not be much of an improvement if all it did was put human agents at the mercy of "probabilism" rather than determinism. Are human choices to depend on random branch selection, operating over different-sized sets of alternatives? This does not sound like a satisfactory basis for controlled human action, arising from deliberative reasoning. If the choices we make are under our control, they have to be to be grounded on something other than branch attrition.

[6] McCall (1987), (1999), (2008), and McCall and Lowe (2005).

[7] Branching spacetime does in fact allow for particular instances of 100% determinism, in cases where all the branches above an A-node are B-branches. E.g., in all instances where the two ends of a copper wire are connected to a battery, current flows in the wire.

6. Control

What exactly does it mean to say that, most of the time, we "control" our actions? A skilled golfer can play a "controlled slice", while a beginner can be plagued with uncontrolled slices. An emotional person may weep uncontrolledly; a reserved person tightly controls any display of emotion. Interestingly, much human behaviour falls into the category of being both controlled and indeterministic.[8] People walking in opposite directions down a crowded street, for example, rarely if ever bump into one another, but even a Laplacian Demon could not predict the exact path that any individual chooses to follow. Brownian movement, exhibited by the random motions of molecules in a gas, differs in being indeterministic and *un*controlled. Small bumps and irregularities in the snow make a skier's precise trajectory indeterministic, but each turn is beautifully controlled. As every skier knows, there is a world of difference between skiing in control and skiing out of control. Another indeterministic example, in this case highly rational and highly controlled, is playing chess. Chess-playing is an indeterministic process because the moves made by one's opponent are unpredictable. If they were predictable, chess would lose much of its allure.

7. Rational, Controlled, Indeterministic Processes

The vital step in understanding how deliberative action is rational, controlled, and indeterministic, as opposed to probabilistic, i.e. based on random branch selection over sets of possible futures, lies in recognizing the mental element in action. If a human being consisted solely of a probabilistically functioning neural mechanism directing the movements of a material body, then indeed there would be nothing to deliberative choice above and beyond the random selection of an "actual" bodily movement from among different-sized sets of future alternatives. But in addition to brains, human beings have minds. The branching spacetime model of the world, and the definition of what it is for the future to be "open", are described exclusively in the 3rd-person mode, that is, as a scientist would describe an atom, or a whale. But our direct experience of the world, and of ourselves, and how we interact with the world, is conceived in 1st-person terms, not 3rd-person terms. In thinking, dreaming, desiring, intending, planning, and deciding what to do we adopt a subjective stance, not an objective one. Human choices are essentially personal, subjective, and mental. This does not prevent their consequences from being far-reaching and

---

[8] McCall (2009), pp. 146–48; See Chapter 8 above.

objective. It is the subjective, mental element makes deliberation rational and controlled, in addition to being indeterministic.

Without indeterminism, there would be no such thing as deliberation. As a matter of logic, one cannot deliberate and decide what to do if there is only one course of action open. As Richard Taylor remarks (1964), one can *seemingly* deliberate and decide to take a late train home, under the mistaken belief that the trains are still running. One can *seemingly* decide (and try) to move one's leg, unaware that the spinal anesthetic administered some hours earlier in the operating room has not yet worn off. In the train case, one can start the process by going to the station, i.e. one can *initiate* the implementation of the decision to take a late train. But one can't *fully* implement it. In the leg case, one can't even initiate it. Unbeknownst to you, the alternative you chose was not open, did not exist. But these are pathological cases, in which two different alternatives seem to be open, but in reality only one is. The standard instances of practical deliberation logically require that there be at least two alternative, open courses of action, and this implies indeterminism.

The mental element enters into practical deliberation at every stage. First, the deliberator must be *aware* of the existence of different possible options. Second, during evaluation, the assignment of appropriate positive and negative weights to the options is a judgemental activity of the mind, involving the use of practical reason. The process is a rational one, taking into account the strength of possibly conflicting impulses and desires, balancing short-term vs. long-term considerations, and influenced by the probable impact of one's actions on others. ("Why did you even consider putting in a side order of clams when they always make you sick?" "Because I knew George was crazy about them and would lick the platter clean.") Third, the final choice at the end of the deliberative process can be a "difficult" one, meaning that its consequences may be grave, or that two different options are equally balanced. Buridan's Ass starved to death, equidistant between two equally tempting piles of hay. A more intelligent being would flip a coin, or simply make the kind of arbitrary choice we make at the supermarket in choosing one of a hundred identical cans of tomato soup. In all cases, the subjective role of the mind in decision-making is critical.

John Searle points out a characteristic picture of the subjective, 1st-person stance, one that sets it apart from 3rd-person affirmations. This is, that in the realm of subjectivity, the distinction between appearance and reality no longer holds (see Searle (2004), p.85). There is, for example, no difference between saying "I am in pain", and "It seems to me that I am in pain". In the case of practical deliberation, a strong subjective element characterizes both the evaluation of alternative courses of action, and the eventual choice. Weighing options is not like weighing sugar. Options do not come with ready-made weights. Before one option can be weighed against another it must be *weighted*, and the weighting process is a subjective one. One who *seems* to attach more weight to comfort than to convenience in traveling, *really does* weight comfort over convenience. It is in the assignment of weights to options that deliberators exercise the first dimension of their control over the deliberative process. The last dimension emerges in the final choice.

Like the weighting of options, choice is a subjective phenomenon. In the world of the 1st-person, to *seem* to choose, *is* to choose. To *seem* to be in pain, *is* to be in pain. In order to appreciate the crucial role that subjective choice plays in decision-making, consider the difference between human decisions, in which the mind plays an essential role, and the decisions made by a probabilistically functioning neural mechanism. As was described above, probabilistic behavior in the objective 3rd-person world is based on global random selection of an actual future from among sets of possible futures, these sets being of different relative proportions. If "persons" are merely neural mechanisms, then their "decisions" are modeled on this pattern. But in fact, a choice made by a human deliberator is very different. It is a mental event, not caused by or supervenient upon random branch selection in the physical world, but itself the cause of branch selection, and the accompanying bodily movement. A mental choice, with physical effects, is a paradigmatic example of mental causation. (On mental causation, see Kim (1998)). Significantly, this is not a case of causal overdetermination, since the physical event that the mental event causes, i.e. the bodily movement, does not already possess a physical cause. This needs to be made clearer.

8. Indeterminism in the Brain

As was stressed above, without indeterminism, without at least two different possible optional actions stemming from the same set of initial conditions, there would be no such thing as practical deliberation. Suppose someone, X, is deliberating about whether to do A, B, or C. Throughout the entire deliberative process, these options remain "open", meaning that X can make a bodily movement initiating the implementation of any one of them at any time. In X's brain there are motor neurons, the activation of any one of which will lead to implementation. Let n(A) be the neuron, the activation of which initiates the implementation of A, and similarly for n(B) and n(C). Before X reaches a decision, each motor neuron is potentially activat*able*, but it is undetermined which one will be actually activat*ed.* It is the indeterministic functioning of X's brain that keeps the options A, B and C open. When X reaches a decision, and chooses one of the three options, say option B, what causes the activation of motor neuron n(B)? We have already examined, and rejected, the hypothesis that n(B) is activated by random branch selection in the spacetime model. There remains only one possibility, that the activation of n(B) is caused neither randomly nor physically, but by the mental, phenomenological event that we call "X's choice". Despite the misgivings of generations of philosophers since the time of Hobbes and Descartes, it would seem that mental causation plays an essential role in deliberation, decision, and action. One of the principal objections to invoking a mental cause in this context, namely the easily made philosophical assumption that every physical effect must have a physical cause, is vitiated in this case by the fact that, because of the indeterministic functioning of the brain, the activation of the motor neuron n(B) has *no* physical cause. Instead it has a mental cause. Having no physical cause, the activation of n(B) is not causally overdetermined.

9. How Rational Agents Exploit Neural Indeterminism

Can the overall indeterministic functioning of the billions of neurons in the human brain be given a naturalistic explanation? Does such functioning have "survival value"? It seems logical to suppose that it does, and that the role of "keeping one's options open", played by neural indeterminism, was one of the most important factors in human evolution. Compare a human brain that behaves indeterministically, with a deterministic neural mechanism functioning on a "stimulus/response" basis. When confronted with a challenge, whether for living space, or food, or physical combat, a creature with a one-option brain would seem to be at a disadvantage compared to a creature with a multi-option brain. An essential component of human rationality is *practical reason*, the ability to examine different courses of action and select the best. Without multiple options, generated by neural indeterminism, an agent is incapable of reasoning in a practical way. Of course, practical reasoning comes with a risk, namely that one may choose the wrong option, a course of action unsuited to the prevailing situation. But human evolution demonstrates, I think, that living riskily brings higher rewards, and a higher level of development, than living safely and predictably.

A long time ago, Plato and Aristotle differed on the question of the Good, and its relationship to human action. Plato, the idealist, maintained that all of us, at all times, seek the good in the sense of acting in accordance with what we believe to be best. If we behave badly, it is through ignorance of where our true good lies. Aristotle, the realist, believed that it was possible to know what is best, but not to do it. Of the two philosophers, Aristotle seems to be closest to recognizing not only the judgemental role of practical reason, but also the possibility that ultimate choice may not accord with deliberative evaluation, with "what one thinks is best". Aristotle judged that this was evidence of human *akrasia*—"weakness of will". This may or may not be so, but the possibility of "knowing the good and not doing it" seems an inevitable consequence of the power of deliberative free choice.

References

Belnap, N.D. "Branching Space-Time", *Synthese* 93, pp. 385–434.

Belnap, N.D, Perloff, M. and Ming Xu (2001), *Facing the Future*, New York: Oxford University Press.

Grünbaum, A. (1963), *Philosophical Problems of Space and Time*, New York: Knopf.

Heisenberg, W. (1958), *Physics and Philosophy*, New York: Harper & Row.

Kim, J. (1998), *Mind in a Physical World*, Cambridge, MA: MIT Press.

McCall, S. (1969), "Time and the Physical Modalities", *The Monist* 55, pp. 426–446; reprinted in Freeman and Sellars (eds.) (1971) *Basic Issues in the Philosophy of Time*, pp. 102–122. La Salle, IL: Open Court.

—— (1976), "Objective Time Flow", *Philosophy of Science* 43, pp. 337–362.

—— (1984), "A Dynamic Model of Temporal Becoming", *Analysis* 44, pp. 172–176.

—— (1987), "Decision", *Canadian Journal of Philosophy* 17, pp. 261–288.

—— (1994), *A Model of the Universe*, New York: Oxford University Press.

—— (1997), "Time Flow, Non-Locality, and Measurement in Quantum Mechanics", in S. Savitt (ed.), *Time's Arrows Today*, pp. 155–172. New York: Cambridge University Press.

—— (1999), "Deliberation Reasons and Explanation Reasons", in Jackendoff, Bloom and Wynn (eds), *Language, Logic, and Concepts*, pp. 97–108. Cambridge, MA: MIT Press.

—— (2000a), "QM and STR: The Combining of Quantum Mechanics and Relativity Theory", *Philosophy of Science (Proceedings)* 67, pp. S535–S548. Chapter 18 of this volume.

—— (2000b), "Towards a Lorentz-Invariant Theory of Collapse", in M.C. Duffy and M. Wegener (eds), *Physical Interpretations of Relativity Theory*, pp. 111–122. Sunderland; British Society for the Philosophy of Science.

—— (2008), "How to Make a Decision", in D. Vanderveken and D. Fisette (eds), *Actions, Rationality and Decision*, pp. 325–338. London: College Publications. Chapter 4 of this volume.

—— (2009), contribution to J.H. Aguilar and A.A. Buckareff (eds), *Philosophy of Action: 5 Questions*, pp. 141–154. Copenhagen: Automatic Press/VIP.

—— (forthcoming), "Controlled Indeterministic Processes in Action Theory", in D. Vanderveken (ed.), *Attitudes and Action in Discourse*. Also on http://www.ucl.ac.uk/~uctytho/. Chapter 8 of this volume.

McCall, S. and E.J. Lowe (2005), "Indeterminist Free Will", *Philosophy and Phenomenological Research* 70, pp. 681–690.

Searle, J. (2004), *Mind: A Brief Introduction*, New York: Oxford University Press.

Taylor, R. (1964), "Deliberation and Foreknowledge", *American Philosophical Quarterly* 1, pp. 73–80.

{ 11 }

Does the Brain Lead the Mind?

Over the last 25 years, experimental findings published by Benjamin Libet have indicated that conscious acts of will are preceded by a characteristic kind of brain event of which the agent is not conscious. It, Libet says, rather than the will, is what causes actions. His discoveries, if correct, would seem to imply that the notion of a free, conscious will is an illusion, and that actions are initiated by neural processes not under conscious control. In what follows it is argued that Libet's conclusion is incorrect, and that other evidence points to the essential causal role of consciousness in voluntary action.

In Libet's experiment, subjects sat at a table in front of an oscilloscope "clock" on which a spot of light revolves around a circular dial divided into 60 sections (Libet (1999), p.50). The spot revolves once every 2.56 seconds. Subjects were asked to perform a simple flick of the wrist whenever they wanted, and to report the oscilloscope "clock time" W at which they were first aware of the wish or urge to perform the action. The oscilloscope clock had to be much faster than an ordinary clock, in order to show time differences of the order of a few milliseconds.

While the subjects were moving their wrists, and recording the times of their urges to do so, an EEG machine with electrodes attached to their scalp recorded brain activity, while an EMG (electromyogram) detected the exact activation time of the arm muscle above the wrist. When plotted on a single time chart, these various data revealed that the conscious wish or urge to move the wrist occurred about 200 ms. before muscle activation.[1] But the EEG also provided the unexpected result that a spike of cerebral neural activity, the "readiness potential" RP, occurred much earlier: 550 ms. before muscle activation and 350 ms. before W, the time of the conscious wish. In Libet's experiments, readiness potentials are not recorded in the absence of a conscious wish. Nor do the data

[1] Critics have questioned the accuracy of Libet's method of timing the occurrence of W. Libet has replied at length: see in particular his BBS paper (1985), p. 532 and pp. 534–35.

show them as occurring when a conscious urge to move the wrist is succeeded by an immediate conscious veto of that act after it is willed.[2] It would appear therefore that the unconscious physical readiness potential RP, rather than the conscious urge following it, is the true cause of action. In Libet's words, "the brain…'decides' to initiate or…prepare to initiate the act…before there is any reportable subjective awareness that such a decision has taken place."[3] It would seem, accordingly, that the mind follows the brain, rather than the other way around.

Libet's conclusions have been welcomed by physicalists and materialists, who deny either that consciousness exists at all, or that it plays any causal role in the physical world.[4] They have been disputed by others, notably by Mele, who makes the valid point that Libet's "readiness potential", occurring at 550 ms. before the time of the action A, should in no way be described as a *decision* or an *intention* to perform A. It could be described, Mele says, as an (unconscious) *urge* to flex the wrist, but as Mele points out, "having an urge" to do something is a long way from forming an explicit *intention* to do it, still less *deciding* to do it. In Mele's words:

> Nothing justifies the claim that what a subject becomes aware of at time W [the time at which subjects reported the "first awareness of the wish to act"] is a *decision* to flex that has already been made or an *intention* to flex that has already been acquired, as opposed, for example, to an *urge* to flex that has already arisen. (Mele (2006), p.40)

Mele's conclusion is that despite what Libet says, his experimental results leave it open that the subjects' flexing of their wrists may have been caused by an act of conscious will. In what follows I consider a different but similar situation, a real-life example in which the causal role of conscious will seems undeniable.

Imagine a heat of the 100 metre dash at the Olympic Games, an event where, as in Libet's experiment, every millisecond counts. The runners line up at the start. At "ready" they kneel in the starting blocks, at "set" they lift their knees from the track, and when the starter's pistol is fired they start. Unlike Libet's subjects they do not move their muscles any time they want, but strive to activate them, with explosive force, as soon as possible after the gun. Does conscious will play a key causal role in this process? It would appear so: at least, no other causal explanation seems attractive. I consider three possibilities.

---

[2] "In the absence of the muscle's electrical signal when being activated, there was no trigger to initiate the computer's recording of any RP that may have preceded the veto; thus, there were no *recorded* RPs with a vetoed intention to act." (Libet (1999), p. 52)

[3] Libet, Gleason et al., in a 1983 contribution to the journal *Brain*, quoted in Al Mele, *Free Will and Luck* (2006), p. 33.

[4] Daniel Dennett's *Consciousness Explained* (1991), and *Freedom Evolves* (2003), contain discussions of Libet's results.

(1) A follower of Libet might suggest that EEG monitoring of the runners' brain activity would reveal an unconscious readiness potential RP in each case, occurring roughly 550 ms. before the initial muscle activation at the start. This RP would cause the runner to start running. But that seems highly implausible. How would a runner's brain be able to anticipate the firing of the starting pistol by 550 ms.? If the pistol failed to go off, would the runners start regardless?

(2) More plausible, perhaps, is the hypothesis that EEG monitoring would reveal the existence of a readiness potential RP in every runner's brain, but that the time of initiation of each RP before the start varied. Some RP's might commence a minute before the start, others only a few milliseconds. This I suppose is possible, but the absence of any fixed time at which a runner's RP would initiate muscle activation prevents hypothesis (2) from providing a true causal explanation. Why would such an RP initiate movement immediately after the gun, rather than before it, or at some other time?

(3) An alternative explanation of what happens at the starting line is that starting is a stimulus-response situation. When the runners hear the gun, they start. No conscious act of will is required. The runners with faster responses get ahead of those with slower responses. This explanation eliminates conscious will and conscious decision entirely, and no doubt would appeal to physicalists. But it cannot be right, for if it were there would be no false starts. False starts, in which the runner starts before the gun, are a recurrent nightmare for racers. The rules are very clear, two false starts and you're out. The stimulus-response explanation cannot explain false starts; it seems not unreasonable to think they are caused by the over-excited conscious state of the racer getting ahead of itself and issuing the command to start before it hears the gun.

The latter may sound like an over-anthropomorphized description of what is going on, and there may exist other theories of what false starts consist in. But I can't conceive of a theory of false starts that makes no essential reference to consciousness and acts of will. One could, perhaps, imagine a power of veto that the mind exercises over bodily movement until the instant the gun sounds. At that instant the veto is lifted, and the muscles contract. But the question still arises, what lifts the veto? More significantly, what lifts the veto in some cases *before the gun fires*, resulting in a false start?

In view of (1)–(3) above, it would appear that the significance of Libet's readiness potential must be re-assessed. Instead of being the *cause* of the subjects' decision to move their wrists, I suggest it be understood as a *necessary condition* of the decision.[5] If a readiness potential were detected in the case

---

[5] Mele (2009), p. 81, makes this suggestion. He points out that (1) "whenever you wiggle your finger, signal S appears a second before you wiggle it" does not entail (2) "whenever signal

of competitors in the hundred meter dash, it could be interpreted as a necessary condition of their ability to start as soon as possible after the starting pistol: not a causally sufficient condition, but a necessary condition. Over and above the necessary condition of the readiness potential, which is physiological and unconscious, a conscious decision in the case of the starting pistol example would be needed for a full-blooded cause. Given this reinterpretation of the readiness potential, the natural thing to say is that Libet's experiment, and the starting pistol example, are instances of the mind leading the brain rather than vice versa.

References

Dennett, D.C. (1991), *Consciousness Explained,* New York: Little Brown.

—— (2003), *Freedom Evolves,* New York: Penguin.

Libet, B., C.A. Gleason, E.W. Wright and D.K. Pearl (1983), "Time of conscious intention to act in relation to onset of cerebral activities (readiness-potential); the unconscious initiation of a freely voluntary act", *Brain 106*, pp. 623–42.

Libet, B. (1985), "Unconscious cerebral initiative and the role of conscious will in voluntary action", *The Behavioral and Brain Sciences* 8, together with Open Peer Commentary and Author's Response, pp. 529–66.

—— (1999), "Do we have free will?" in Libet, Freeman and Sutherland (eds), *The Volitional Brain*, pp. 47–57. UK: Imprint Academic.

—— (2004), *Mind Time*, Cambridge, MA: Harvard University Press.

Mele, A. (2006), *Free Will and Luck*, New York: Oxford University Press.

—— (2009), *Effective Intentions*, New York: Oxford University Press.

S appears, you wiggle your finger a second later". Buying a lottery ticket is a *causally necessary* condition of winning a prize, but it is not *causally sufficient*.

{ 12 }

3D/4D Equivalence, the Twins Paradox, and Absolute Time

Storrs McCall and E.J. Lowe

1. Introduction

A lot has been written recently on the subject of whether the physical world is made up of three-dimensional objects which endure through time, or four-dimensional objects which are extended in time. The ordinary, everyday view is that a rabbit, say, is a 3D continuant which changes yet retains its identity from one moment to the next. Opposed to this is the theory that a rabbit is a 4D object composed of, or divisible into, temporal segments or parts. A number of considerations favour the second view, among them the fact that it is impossible to understand or come to terms with puzzles such as the train/tunnel and the twins paradoxes without recourse to four-dimensional geometry. Is a choice then forced upon us between 3D and 4D ontology, between common sense and science? Not in our opinion. The position of this chapter is that the 3D and the 4D descriptions of the world are equivalent, in a clear and precise sense to be explained, and that it is not a question of one being true and the other false. In describing a material object in 3D or in 4D terms we are giving alternative descriptions of one and the same thing, and to see the world aright is to realize this.

Our second purpose is to examine what theory of time sits best with the thesis of 3D/4D equivalence, and to this end a study of the twins paradox yields valuable insights. When twin *S* steps off her spaceship she is markedly younger than twin *T*, and their difference in age is not something that can be made to disappear by a Lorentz transformation. The principal key to understanding the paradox is, as in the case of the train/tunnel paradox, four-dimensional geometry. But it is unclear that four-dimensional geometry alone explains differential ageing—the wrinkles on the face of one twin and the smooth skin of the other.

Clocks measure temporal length to be sure, but we shall argue that they also do something else, namely measure the A-theory phenomenon of temporal passage. Since the two twins are, in a perfectly good sense, 3D "clocks" which keep time with their heartbeats, and since they agree at the start of the journey and disagree at the end of it, it would appear that time flows not "absolutely" or "globally", but at different rates in different reference frames. This will in fact be the overall conclusion of the chapter. If we lived in a world that was exclusively 4D, then a B-theory of time based on relations of "earlier" and "later" would be sufficient. But since we live in a world that is also (and equivalently) 3D, and since each twin manifestly undergoes temporal passage and ageing in the particular frame or frames she occupies, A-theory conceptions are also indispensable. There is, in the physical world, differential frame-dependent time flow, rather than a universal "tide of becoming", and the wrinkles on the twins' faces bear witness to it. Or so we shall argue.

2. The 3D/4D Debate

Most people take it for granted that familiar 3D objects change, being qualitatively *different* at different times, while also continuing in existence, i.e. remaining numerically *identical* from one moment to the next. The coffee cup I drink out of is cracked; last month it wasn't, yet it is the same cup I have used for the past three years. Is there a problem about this? Aristotle remarks that a man may be pale at one time and dark at another, and in this sense "receive contraries" (*Categories* 4a10–21). Beyond saying that "nothing like this is to be seen in any other case", Aristotle does not appear to be troubled by this circumstance, but takes it as part of what it is to be an individual substance *X* that different properties may be predicated of *X* at different times.

In his essay "Identity, ostension, and hypostasis" (Quine 1953), Quine gave a Heraclitean answer to Aristotle's problem about change that has set the tone for all subsequent discussions of enduring objects and identity through time. How is it possible to step twice into the same river, e.g. the river Cayster? Quine's answer is that because a river is an entity continuously in flux, more like a process than a thing, strictly speaking what is stepped into is a river-stage rather than a river. A river-stage is a momentary aggregation of all the water molecules which lie between the river's banks at a certain time. It is one of a huge multiplicity of temporal stages or parts which collectively constitute the 4D object known as River Cayster. Since fresh waters are always flowing upon us, there is no enduring 3D object into which we can step even once, let alone twice. Echoing Hume, Quine asserts that only the relation of resemblance, not that of identity, holds between river-stages. Quine calls this relation "river-kinship". He agrees with Hume that in imagining the existence of an enduring 3D river, the resemblance in question causes us to "substitute the notion of identity, instead of that of related objects" (*Treatise* I, pt 4, sect. 6).

Over the years, the proposal to replace the 3D ontology of enduring objects having no temporal parts, by a 4D ontology of collections of temporal parts which do not endure, has generated a large literature. On the 4D side: Quine 1953, 1960, 1981; Smart 1972; Perry 1972: 466–69; Armstrong 1980; Lewis 1983: 76–77, 1986: 202–4, 2002; Heller 1984, 1990; Sider 1997, 2001; Le Poidevin 2000. Four-dimensionalists are divided into (i) those for whom the basic ontological elements are 4D volumes (the "worm" ontology), and (ii) those who consider temporal stages to be basic, which are either "thin" 4D objects with nonzero temporal extension, or are instantaneous 3D slices of 4D volumes (the "stage" ontology). (Instantaneous slices of 4D objects, though three-dimensional, of course differ sharply from enduring 3D objects.) Those who support enduring 3D objects include Geach 1965/1972; Chisholm 1976: appendix A; van Inwagen 1981, 1990; Mellor 1981: 104; Thomson 1983; Lowe 1987, 1998: 114–25; Simons 1987, 2000; Haslanger 1989, 1994. Hawley 2001 surveys the debate.

3. Four-Dimensionalism and Relativity Theory

The train/tunnel paradox raises the issue of four-dimensionalism in a particularly direct way. A train, when standing in a tunnel, fits exactly inside it. If the tunnel is equipped with doors which open and shut rapidly, the two ends of the train make light contact with the doors when closed. Imagine the train moving rapidly through the tunnel, and let E be the event of the centre of the train coinciding with the tunnel's mid-point. At that moment two observers with stop-watches, one at each end of the tunnel, will observe the two ends of the train safely inside, with a certain margin to spare. Therefore, in the tunnel's rest-frame, the train is shorter than the tunnel. This apparent shortening of the train is not an illusion, but perfectly real and objective. If the tunnel doors work quickly enough, there will be a short interval during which they are closed and the train is inside. A photographer in a balloon, stationed vertically over the mid-point of the tunnel, will be able to take a photo showing both doors shut.

In the train's rest-frame, on the other hand, there is a different story. Two observers at each end of the train, with synchronized stop-watches, will record at a time simultaneous with E that the doors are open and both ends of the train are sticking out of the tunnel. Again this is no illusion. A second photographer in an airplane, flying above the tunnel at the same velocity as the train, will take a photo showing the doors open and the two ends of the train plainly visible. In the train's rest-frame, the train is longer than the tunnel, not shorter than it. If some matter of great importance hung on the question of whether the tunnel or the train was longer, and if you were the judge in a trial at which the balloon-photo and the airplane-photo were produced, which would you choose?

A satisfactory resolution and understanding of the train/tunnel paradox cannot be had without recourse to four-dimensional geometry. Diagrams which represent the train and the tunnel as 4D objects in their respective inertial frames are found in Maudlin 1994: 54 and Balashov 2000: 336. These diagrams show clearly how it comes about that the train is both longer than the tunnel in one frame, and shorter than it in another. The length contraction of the train in the tunnel-frame, and of the tunnel in the train-frame, are best understood as "perspectival" phenomena produced by the structure of relativistic spacetime geometry. Analogously, the foreshortening of distant objects in the human visual field, and the apparent trapezoidal shape of rectangular objects perceived obliquely, are best understood as perspectival features of 3D geometry: witness the perspectival analysis of the variable length of Cyrano de Bergerac's nose in Janssen and Balashov (forthcoming). Since to grasp the train/tunnel paradox and similar relativistic effects requires an understanding of four-dimensional geometry, the paradox supplies a strong argument in favour of a 4D as opposed to a 3D ontology.

This being said, it is noteworthy that the Special Theory of Relativity was originally presented by Einstein in 1905 in three-dimensional style, with 3D measuring rods and clocks suffering contractions and dilations when observed in different reference frames. Only after the appearance of Minkowski's paper in 1910 did Einstein adopt the 4D spacetime interpretation of relativity theory. In this paper, he notes, Lorentz transformations are nothing but rotations of the coordinate system in 4D space (Schilpp 1949: 59, Craig 2001: 78–79). Craig remarks that "the seriousness with which Einstein took [the four-dimensional] conception may be seen in the fact that when his life-long friend Michael Besso died, Einstein sought to comfort his bereaved family by reminding them that for physicists Besso had not ceased to exist but exists tenselessly as a permanent feature of the spacetime reality" (p. 79).

4. The Thesis of 3D/4D Equivalence

If common sense recommends an ontology of enduring 3D objects, and if relativity theory is best understood through the medium of 4D objects at rest in different inertial frames, are we forced to choose between the two competing ontologies? No. The thesis of 3D/4D equivalence states that objects of the physical world can be described using either three-dimensional or four-dimensional language, and that the descriptions are equivalent in the sense of being intertranslatable. Furthermore, there is no "fact of the matter" in the world which makes one of the descriptions true and the other false. Some examples drawn from different philosophical and mathematical contexts will help make the equivalence thesis clear.

The first example concerns circles. How is the concept of a circle defined? There are two standard ways, "statically", as the set of all points equidistant from a given point, or "dynamically", as the locus of a point moving continuously at a fixed distance from a given point. The two definitions are equivalent. One and the same object is described by them. Just as a circle can be defined either "statically" or "dynamically", so a physical object like a rabbit can be defined either in 3D language, as an active animal that hops about, or in 4D language, as a set of rabbit-stages. Alternatively, in 4D mode, a rabbit is an elongated volume of several years' temporal extension, early cross-sections of which yield a baby rabbit nursing its mother, and later cross-sections a fully developed adult.

The second example derives from A.N. Whitehead's method of extensive abstraction. In 1920: Ch. 4, Whitehead defines the notion of a "point" in space to be an infinitely descending nested set of volumes. An alternative approach in topology is to define a "volume" as a connected set of points. In giving theoretical foundations for topology or geometry, we may (i) define "point" in terms of "volume", or (ii) define "volume" in terms of "point". Since, given Whitehead's insight, the two methods yield the same results, a conceptual structure which takes "point" as primitive is equivalent to a structure which takes "volume" as primitive.

In the same way, we propose, the argument over whether physical objects are "really" three- or four-dimensional is also pointless (no pun intended). Suppose someone were to raise the ontological question of whether space is "really" composed of points, or "really" composed of tiny volumes. Before Whitehead's definition, it was perhaps arguable that this question raised a substantive philosophical issue. But if a "point" can be defined as an infinite set of volumes, and if, equally, a "volume" can be defined as an infinite set of points, then to argue about which of these items are the real constituents of space is, in a different sense, pointless. In analogous fashion, if physical objects can be considered as three-dimensional entities existing through time, and if they can equally well be considered as four-dimensional objects composed of, or divisible into, temporal parts, then the dispute over whether the physical world is really 3D or 4D is similarly empty. It may be both, or it may be neither, or it may be either, depending on which features of the world we wish to focus on. But it will never be one to the exclusion of the other; never the case that the 3D view is vindicated and the 4D view defeated, or vice versa.

A third example underlines the tendency of ontological disputes to disappear under the weight of a "principle of indifference", according to which it is immaterial which of two underlying theories is true.

Imagine that intelligent life on earth disappears, and that many years later Martians arrive and discover, among other artifacts, quantities of sheet music,

LPs, and compact discs. Although they have no ears and cannot hear anything, the Martians have a keenly developed aesthetic sense and become very intrigued about what "music" might be. A lively debate springs up between those who think that music consists of patterns of notes printed on 5-line staffs, and those who think music consists of cyclical irregularities in the grooves of LPs and compact discs. The debate is spirited, with many articles in learned journals, until one day a particularly intelligent participant produces a translation manual which demonstrates the existence of a complete structural isomorphism between the "sheet music" and the "groove" interpretations. (A small correction factor to the overall isomorphism correlates minor systematic variations in groove patterns with the occurrence of words like "piano" and "violin" on sheet music, but with these corrections the translation is perfect.) Given a sheet music description, a groove pattern description is constructible and vice versa. Once the translational equivalence of the two theories is established, interest in the ontological question of what music really is disappears. Any intelligent Martian knows that whether described in sheet music or in groove pattern terms, Beethoven's Fifth Symphony is a landmark composition which can be studied, loved and appreciated in either of its two ontological manifestations. As with the thesis of sheet music/groove equivalence, so is the case with the thesis of 3D/4D equivalence.

5. The Twins Paradox

Writ large, the twins paradox has twin *S* remaining on earth, while twin *T* flies on a space-ship to Alpha Centauri and back. When she returns she has aged 30 years, while twin *S* has aged 40 years. The difference in elapsed time does not arise from the fact that *T* has suffered accelerations and decelerations, while *S* has remained at rest in something approximating an inertial frame. In the 3-clock version of the paradox, let clocks 1 and 2 be synchronized at space-time point *X*. Suppose that each clock follows a different inertial path, and let clock 3 be synchronized with clock 2 at point *Y* (Bondi 1964: 80–87; Marder 1971: 73–78, 112–3). If clock 3 follows a third inertial path, and if this path intersects clock 1's path at *Z*, then the total elapsed time of 2 and 3 will always be less than the elapsed time of 1 (figure 1).

In Minkowski geometry with metric signature + ---, in which the length of a line segment *ds* is given by the formula

$$ds^2 = dt^2 - dx^2 - dy^2 - dz^2$$

the proper elapsed time *dt* of a clock at rest in an inertial frame equals the length *ds* of the corresponding line segment. Consequently, if we build a geometry out of inertial lines, i.e. the paths followed by objects at rest in inertial

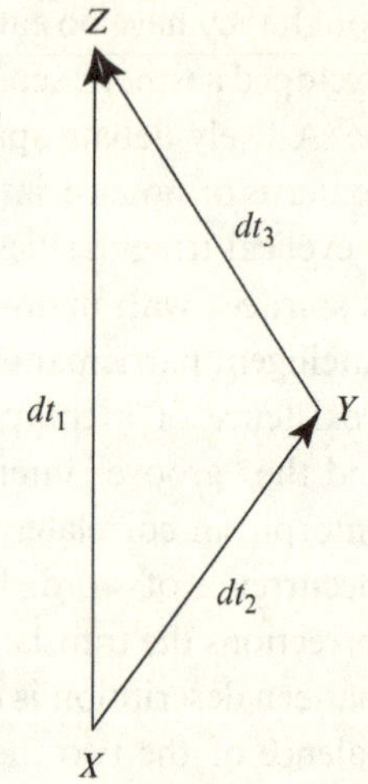

FIGURE 1

frames, the lengths of which are equal to the objects' proper times, we arrive at a simple inequality. What the clock paradox rests on is that in such a geometry the length dt_1 in figure 1 is always greater than the sum of the lengths dt_2 and dt_3:

$$dt_1 > dt_2 + dt_3$$

In Euclidean geometry, triangle inequality operates the other way around: $dt_1 < dt_2 + dt_3$. But in Minkowski inertial line geometry, one side of a triangle is greater, not less, than the sum of the other two sides. In relativistic athletics, the medal for the shortest elapsed time goes not to the runner who runs straight, but to the runner who takes a zig-zag path to the finish. What lies at the root of the twins paradox is a reversed law of triangle inequality.

The twins paradox carries important consequences for our everyday notion of time. If *S* stands at the sideboard while *T* walks over to the table and back, *T* will have aged less than *S* when she returns. The way to keep youthful is to keep on the move! As in the train/tunnel paradox, the difference in age between stay-at-home *S* and traveller *T* is not an illusion, but is perfectly real. In fact it is even more real than the difference in length of train and tunnel, because the two photographs showing the latter were shot in different inertial frames, whereas the twins' difference in age is witnessed by increased grey hair and wrinkles once reunited. These differences are absolute, Lorentz-invariant, and cannot be transformed away by changing reference frames. We shall argue that the differential ageing of the twins in different frames, when studied within the context of the A-theory of time, requires that the idea of global or absolute becoming be abandoned in favour of *frame-dependent* temporal passage.

Consider in more detail the two individuals *S* and *T* who stand together at the sideboard, separate when *T* walks to the table, and then are re-united. At one level of explanation, the fact that *S*'s age has increased relative to *T*'s is explained by the fact that in Minkowski geometry the unbroken path followed by *S* is longer than the broken path followed by *T*. But there is another level of explanation, also relativistic, which concerns not the temporal *lengths* of the sides of the triangle in figure 1 but the temporal *rates* at which physical processes take place. The distinction between temporal lengths and temporal rates takes us back to 3D/4D equivalence. The triangle of figure 1 which measures elapsed time is a triangle in 4D geometry, whereas the clock or the individual which follows the 4D trajectory and undergoes temporal passage is necessarily a 3D object. Only for such objects do the notions of "temporal process" and "rate of temporal process" make sense. However, as will be seen, 3D/4D equivalence ensures that the A-theory concepts of "process" and "rate", and the B-theory concept of "temporal distance", are two sides of the same coin.

Three-dimensional objects are like little clocks. Their 4D world-lines, divided into equal segments like a ruler, measure quantity of elapsed time. But *qua* 3D objects they also measure rates of elapsed time, which are A-quantities. For traveller *T*, who moves to the table and back, time passes more slowly than for *S*. While *T* experiences *x* heartbeats, undergoes *t* cycles of her neural alpha-rhythym, and hums *z* bars of *Tosca, S* experiences $x + dx$ heartbeats, $y + dy$ neural cycles, and hums $z + dz$ bars. It would be tempting to say that *S* has crammed more experiences and events than *T* into the same time period, but this would be incorrect. The time periods themselves are unequal. Between their initial parting and later reunion, *S* has more time available than *T*. Consequently, for him, time moves more quickly, while for *T* it moves more slowly. Statistically, using counts of heartbeats or neural cycles per week, it should be possible to confirm that office workers who are confined to their desks age more rapidly than mail deliverers and taxi-drivers, who change their inertial frames many times a minute.

This conclusion is admittedly controversial. It violates our instinctive belief that (by definition?) time always moves at the constant rate of one second per second (Prior 1968: 2–3), and that variations in this rate are unthinkable. For this reason it is worth going over once more the argument for variable rates of temporal flow, this time focussing on inertial frames rather than lengths of world-lines.

A significant difference between *S* and *T* is that *T* successively occupies (is at rest in) two different inertial frames f_2 and f_3, corresponding to two of the sides of the triangle of figure 1, while *S* occupies only one frame, f_1. Since a similar elapsed time discrepancy is observed for any two objects, one of which successively occupies two different inertial frames and the other only one, the conclusion is strongly indicated that the time difference is due to differential flow rates in different frames. And this in turn implies that global or absolute time flow, and consequently Global or Absolute Time, is non-existent. Those who argue that nothing in the theory of relativity rules out absolute time (Tooley 1997: 338–54; Craig 2001), cannot reconcile this position with the phenomenon of differential ageing in the twins paradox.

Unlike such Lorentz-invariant quantities as the B-theory notion of spatio-temporal length, the A-concept of temporal passage or flow is inherently frame-dependent, and its rate is particular to each and every inertial frame. Therefore there can be no such thing as Absolute Time. This is admittedly a large conclusion to derive from the small beginnings of the twins paradox, conjoined with 3D/4D equivalence, but we see no alternative to it.

References

Armstrong, D.M. 1980. Identity through time. In *Time and Cause*, ed. van Inwagen, 67–78. Boston: Kluwer.

Balashov, Y. 2000. Persistence and space-time. *The Monist* 83: 321–40.

Bondi, H. 1964. *Relativity and Common Sense*. New York: Anchor Books.

Chisholm, R.M. 1976. *Person and Object.* LaSalle, IL: Open Court Pub. Co.

Craig, W.L. 2001. *Time and the Metaphysics of Relativity.* Boston: Kluwer Academic.

Geach, P.T. 1965/1972. Some problems about time. *Proceedings of the British Academy*, reprinted in his *Logic Matters*, 302–18.

Haslanger, S. 1989. Endurance and temporary intrinsics. *Analysis* 49: 119–25.

—— 1994. Humean supervenience and enduring things. *Aust. J. Phil.* 72: 339–59.

Hawley, K. 2001. *How Things Persist.* New York and Oxford: Oxford University Press.

Heller, M. 1984. Temporal parts of four dimensional objects. *Phil. Studies* 46: 323–34.

—— 1990. *The Ontology of Physical Objects.* New York: Cambridge University Press.

Janssen, M. and Y. Balashov (2003). Review of W.L. Craig, *Time and the Metaphysics of Relativity*, in *British Journal for the Philosophy of Science* 54: 327–46.

Le Poidevin, R. 2000. Continuants and continuity. *The Monist* 8: 381–98.

Lewis, D. 1983. *Philosophical Papers*, vol. 1. New York: Oxford University Press.

—— 1986. *On the Plurality of Worlds.* New York: Blackwell.

—— 2002. Tensing the copula. *Mind* 111: 1–13.

Lowe, E.J. 1987. Lewis on perdurance versus endurance. *Analysis* 47: 152–54.

—— 1998. *The Possibility of Metaphysics.* New York: Oxford University Press.

Marder, L. 1971. *Time and the Space-Traveller.* Philadelphia: University of Pennsylvania Press.

Maudlin, T. 1994. *Quantum Non-Locality and Relativity.* Cambridge, MA: Blackwell.

Mellor, D.H. 1981. *Real Time*. Cambridge: Cambridge University Press.

Perry, J. 1972. Can the self divide? *J.Phil.* 69: 463–88.

Prior, A.N. 1968. *Papers on Time and Tense.* Oxford: Clarendon Press.

Quine, W.V.O. 1953. *From a Logical Point of View.* Cambridge, MA: Harvard University Press.

—— 1960. *Word and Object.* Cambridge, MA: MIT Press.

—— 1981. *Theories and Things.* Cambridge, MA: Harvard University Press.

Schilpp, P.A. ed. 1949. *Albert Einstein: Philosopher-Scientist.* New York: Harper & Row.

Sider, T. 1997. Four-dimensionalism. *Phil. Review* 106: 197–231.

—— 2001. *Four-Dimensionalism: An Ontology of Persistence.* New York and Oxford: Oxford University Press.

Simons, P. 1987. *Parts: A Study in Ontology.* New York: Oxford University Press.

—— 2000. How to exist at a time when you have no temporal parts. *The Monist* 83: 419–36.

Smart, J.J.C. 1972. Space-time and individuals. In *Logic and Art: Essays in Honor of Nelson Goodman*, ed. Rudner and Scheffler, 3–20. Indianapolis: Bobbs-Merrill.

Thomson, J.J. 1983. Parthood and identity across time. *J.Phil.* 80: 201–20.

Tooley, M. 1997. *Time, Tense, and Causation.* Oxford: Clarendon Press.

van Inwagen, P. 1981. The doctrine of arbitrary undetached parts. *Pacific Phil.Quart.* 62: 123–37.

—— 1990. Four-dimensional objects. *Nous* 24: 245–55.

Whitehead, A.N. 1920. *The Concept of Nature.* Cambridge and New York: Cambridge University Press.

{ 13 }

Philosophical Consequences of the Twins Paradox

1. The Paradox

The twins paradox runs as follows. Jack stays on earth while his twin Jill steps into her spaceship, rounds Alpha Centauri at 3/5 the speed of light, and returns. Since at this speed the time dilation factor is 4/5, when she gets back Jack has aged 50 years while she is only 40 years older (Darwin (1957); Resnick (1968, 201–209)). In the special theory of relativity, the difference in ages is explained by the fact that when Jill changes direction and starts to return to earth, her "simultaneity classes"—the sets of events which partition spacetime in the different inertial frames she inhabits—undergo a sudden change (see figure 1).

Jill's journey outward from A to B, and homeward from B to C, each takes 15 years as registered by her clock. The events on earth which are Jill-simultaneous with successive five-year anniversaries on her voyage are marked by the intersection of her different "now" lines with Jack's spacetime trajectory. Jack's age increases are in the left-hand column. When Jill changes her inertial frame at B and heads back to earth, her "now" lines move abruptly upward on Jack's lifeline, leaving a gap between X and Y. Suppose that in her absence, in the space of 50 years, Jack reads *War and Peace* three times. Then Jill can truthfully say, as she journeys outward, "He is now reading *War and Peace* for the first time", and coming home she can say "He is now reading *War and Peace* for the third time". But on the assumption that Jill's turnabout is instantaneous, there is no time at which Jill can say that Jack is reading *War and Peace* for the second time. If Jack gets married during the interval XY, there will be no period on Jill's clock which corresponds to the wedding ceremony. For her, the middle section of Jack's life goes by in a flash, and this explains why there is the difference in their ages. It does not explain the *quantitative* difference, since the gap XY is 18 years and when they re-unite Jack is only ten years older. But it does explain why, in Jill's reference frame, there are "missing"

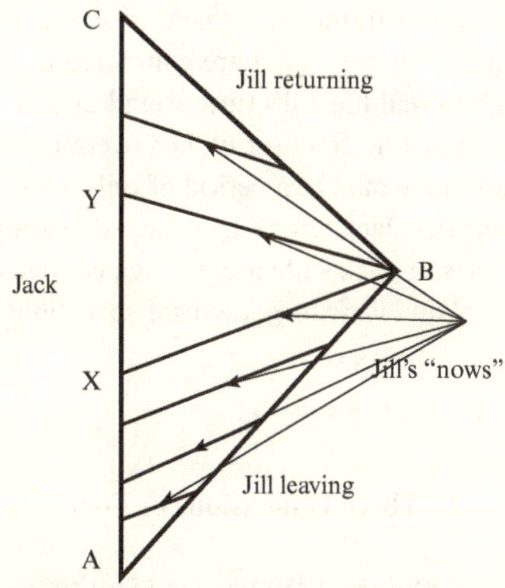

FIGURE 1 *The Twins Paradox*

events in Jack's life. Since Jack inhabits only one inertial frame[1] in contrast to Jill's two, their situations are asymmetric. On Jack's clock, there are no missing events in Jill's life.

Figure 1 provides an intuitive explanation why, in four-dimensional relativistic geometry, the sum of the lengths of two sides of a triangle composed of timelike inertial lines is always less than the third side. (An inertial line is the path of a body at rest in an inertial frame.) Since for any inertial line $dx = dy = dz = 0$, and since

$$ds^2 = c^2dt^2 - dx^2 - dy^2 - dz^2,$$

the length ds of a line segment is equal to the elapsed time cdt. Dividing by the constant c, let the lengths of the segments AC, AB and BC be dt_1, dt_2 and dt_3 respectively. Then figure 1 explains why, in Minkowski geometry, triangle inequality operates in reverse of Euclidean geometry:

$$dt_1 > dt_2 + dt_3.$$

In figure 1, $dt_1 = 50$ years, $dt_2 = dt_3 = 20$ years, and $50 > 20 + 20$. To be sure, in real life Jill's turnaround at point B is not instantaneous, and the triangle ABC will have a curved apex at B. Only in the "three clock paradox" (Bondi (1964, 80–87); Marder (1971, 73–8 and 112–13); McCall and Lowe (2003, 119–20)) will ABC be a perfect triangle. In the three clock paradox clocks 1 and 2, each

[1] If Jack is on earth, he is being continuously accelerated by the pressure on the soles of his feet. A "pure" version of the twins paradox would have Jack floating freely in an inertial frame in space.

travelling in its own inertial frame, are synchronized at A, and clocks 2 and 3 are synchronized at B. When 1 and 3 are compared at C, they are found to disagree. But although in real life Jill's turnaround at B is smooth rather than abrupt, it occupies only a tiny fraction of her overall voyage. Consequently, during the turnaround, there may be a period of only a few hours during which she can truthfully assert that Jack is making a second reading of *War and Peace*. During that period years of Jack's life are, for her, condensed into a matter of minutes. Their loss explains why, in relativistic spacetime, triangle inequality holds in reverse.

2. Three Dimensions or Four?

Since the appearance of Quine's 1950 paper "Identity, ostension, and hypostasis", a protracted, lively debate has taken place between contemporary followers of Heraclitus, and contemporary followers of Parmenides. Heracliteans believe that the world is made up of three-dimensional objects which endure and change in time, while retaining their identity from one moment to the next. Parmenideans, on the other hand, believe that the world is a changeless four-dimensional spacetime continuum, containing material objects which are four-dimensional wormlike volumes extended along the time dimension. Viewed along the time axis, these 4D worms twist, enlarge, shrink, and touch each other so as to perfectly mimic the motions and changes of 3D objects. The ontological question for metaphysicians and philosophers of science is this: is the physical world *really* 3D or 4D?

Philosophers who argue for a 4D ontology include Quine (1950/1953), (1960), (1981); Smart (1972); Perry (1972, 466–69); Armstrong (1980); Lewis (1983, 76–77), (1986, 202–4), (2002); Heller (1984), (1990); Sider (1997), (2001); Le Poidevin (2000). Some 4D supporters prefer a "worm ontology", according to which ordinary objects such as teacups and butterflies are entire elongated 4D volumes. Others consider the basic ontological building blocks to be "temporal parts", instantaneous or thin temporal slices of 4D volumes. For Quine, a purportedly enduring object such as a rabbit is in reality composed of "rabbit stages", short segments of the history of a rabbit. (An instantaneous rabbit slice, though three-dimensional, must plainly be distinguished from a 3D rabbit which endures and changes through time.)

Opposed to the four-dimensional school are those who believe that the physical world is made up of 3D objects which move and interact, are created and destroyed, and retain their identity throughout the time they exist. These are the things Aristotle calls "individual substances" (*Categories*, chapters 2–5). Philosophers who favour a 3D ontology include Geach (1965/1972); Chisholm (1976, appendix A); van Inwagen (1981), (1990); Mellor (1981, 104); Thomson

(1983); Lowe (1987), (1998, 114–25); Simons (1987), (2000); Haslanger (1989), (1994).

The choice between 3D and 4D is sometimes presented as a choice between common sense and science. We can explain the workings of a steam engine in terms of the translations and rotations of 3D pieces of steel, driven by the expansive force of steam. But how are we to explain the famous train/tunnel paradox without recourse to 4D ontology? A train, stationary in a tunnel, fits exactly inside it. But if the train is backed up and run through the tunnel at speed, observers on the ground will record the ends of the train fitting comfortably inside the tunnel at the instant when the centre of the train coincides with the centre of the tunnel. If the ends of the tunnel have glass tops, a helicopter stationed above the tunnel's mid-point will take a photograph showing the entire train inside the tunnel with room to spare. The photo shows the train to be shorter than the tunnel. But at the same instant an airplane, flying above the tunnel with the same velocity as the train, takes a photo showing the two ends of the train protruding from the tunnel. How can there be two photographs, shot from above, one showing the train to be shorter than the tunnel and the other showing it to be longer?

The photograph-version of the train/tunnel paradox presented in the previous paragraph needs fuller discussion, given that the photons travelling upwards to the cameras take time to complete their journey. Imagine that the helicopter contains a movie camera, which is positioned above the midpoint of the tunnel, and which takes numerous shots of the train as it first enters and then exits the tunnel. Let time(a) be the time it takes a photon to travel from the front end of the train to the camera, time(b) the time from the rear end to the camera, time(c) from the tunnel entrance to the camera, and time(d) from the tunnel exit to the camera. Since the tunnel is stationary and the camera is centered, we have, in the tunnel's rest frame, time(c) = time(d) always. There will, however, be only one instant at which time(a) = time(b). That instant, t, will be when the train is positioned symmetrically under the camera, with its midpoint immediately underneath. Before t, time(a) < time(b), and after t, time(a) > time(b).

When the two light rays, or light packets, emitted at t from the train's back and front, converge on the camera accompanied by the "constant" light rays from the two ends of the tunnel, the resulting photo will show the train entirely within the tunnel, with something to spare at each end (as revealed by the glass roof).

Consider now the situation in the rest-frame of the train, where the tunnel is in motion. Assume that the camera in the airplane remains centred above the midpoint of the train, looking down. In the train's rest-frame, we have time(a) = time(b) always, since the time taken for a photon to travel from the front end of the train to the camera in the airplane always equals the time taken for photon to travel from the rear end of the train to the camera. But there will be one instant only at which time(c) = time(d), and when the photons emitted at that instant arrive at the camera they will show the two ends of the train sticking out from the tunnel. The two photos, from two different cameras, show different relative lengths of train and tunnel.

The most obvious and most elegant resolution of the paradox comes by treating the train and the tunnel as 4D objects. In the tunnel's inertial frame, the train is shorter; in the train's frame, the tunnel is shorter. 4D diagrams which make everything clear are found in Maudlin (1994, 54) and Balashov (2000, 336). For many philosophers, the train/tunnel paradox demonstrates that while a 3D ontology suffices for ordinary purposes, true scientific understanding of the world requires four dimensions.

In section 5 below it will be argued that, rightly seen, the choice between 3D and 4D is a matter of "both/and" rather than "either/or". In the next section I shall argue that full and adequate foundations of a theoretical framework in which to situate the twins paradox can be achieved only by recognizing the existence of both 3D and 4D objects. Explanation using 3D elements complements explanation using 4D elements, and vice versa. Neither alone covers every situation: both are needed.

3. The Metrization of Spacetime Using 3D Elements

When Jack and Jill meet after Jill's journey, she has aged 40 years and he has aged 50. What testifies to this? Their clocks, and the wrinkles on Jack's face. In Minkowski spacetime, the length of Jack's world-line, measured by the difference of his time coordinate at the start and end of Jill's journey, is 50 units. To give the answer "50 years" to the question "How long is Jack's world-line between A and C?" presupposes that the region of spacetime which encloses the twins is *metrized*, that it has a metric. If it did not, spacetime would support only an affine geometry, and there would be no answer to the question about Jack's world-line in the last sentence. But, in the metric space enclosing the twins, what philosophical meaning is to be attached to a statement like "Jack wiped his eyes 5 minutes after Jill left"? Does spacetime come equipped with some intrinsic metric, a little pointer that indicates "5 minutes" as the temporal length of that interval? No. The only conceivable entity that can bestow a metric on the time coordinate of Jack's frame is a clock. And a clock is a three-dimensional object. In the same way, the only conceivable entity that can metrize spatial dimensions is a rigid measuring rod, a 3D object that can be transported and applied repeatedly to spatial intervals, giving them a length. Without 3D objects, 4D space would lack a metric.

I can sense criticism building in the face of what many will consider an unacceptably operationalist approach. An obvious rejoinder would be: "The clock does not *give* the temporal interval a length, it merely measures the length that the interval *already has*". Thus Russell, in his debate with Poincaré in the years 1897–1900, rejects Poincaré's basing of the equality of spatial and temporal intervals on an external (and therefore conventional) definition of "congruence". Russell appeals to the idea of an intrinsic spatial metric:

"Whatever one can discover by means of an operation must exist independently of the operation: America existed before Christopher Columbus, and two quantities of the same kind must *be* equal *or* unequal before being measured. Any method of measurement is good or bad according as it yields a result which is true or false. M. Poincaré, on the other hand, holds that measurement creates equality and inequality". (Russell (1899, 687–88), translated in Grünbaum (1963, 44–45))

One can sympathize with Russell's stout realism, but the question of what makes two intervals objectively or intrinsically equal remains unanswered. Riemann, in his inaugural lecture of 1854, had pointed out that if space were discrete or granular, one could determine the equality or inequality of two intervals by counting. (See Grünbaum (1963, chapter 1) for a detailed discussion of these issues.) Such a method would yield an objective, intrinsic result. But since physical space as far as we can determine is continuous, not discrete, the method of determining equality of intervals cannot proceed by counting, but must be by way of measurement. For Riemann, "Measuring consists in superposition of the magnitudes to be compared; for measurement there is requisite some means of carrying forward one magnitude as a measure for the other." (Riemann (1854/1959, 413)). Riemann's words "superposition" and "carrying forward" clearly imply the use of a ruler, i.e. a rigid body which establishes the congruence of two intervals by (i) occupying the first, then (ii) being carried forward, and finally (iii) occupying the second. Without the availability of a three-dimensional transportable measuring device, four-dimensional space-time intervals (which are not transportable) cannot be said to be equal, i.e. will not *be* equal. For this reason a purely 4D ontology is ultimately untenable. Without 3D measuring rods and clocks, no adequate philosophical account of the notion of spatial or temporal congruence, and consequently of spatial or temporal length, is possible.

To this it might be objected, and indeed *has* been objected by at least one colleague, that I have no warrant for assuming that clocks and rods must be 3D objects. Why could they not be 4D? In fact my colleague thinks they *are* 4D. He asks, how could they be used to measure 4D intervals if they were not? This objection goes to the heart of what I am saying about the necessity of complementing a 4D description of the world with 3D elements. What I am saying is that it is impossible to give a clear philosophical account of *measuring* intervals, or of two intervals being *congruent*, or for that matter being of *different lengths*, without the concept of a 3D rod or a 3D clock. The essential thing about a rod, or a clock, is that it must be the *same* rod or clock from one moment to the next, i.e. when we apply it first to one interval, and then to another. In addition rods must remain the same length, and clocks must tick at the same rate, under conditions of slow transport. That's what measuring is all about. Riemann had it right here. For instance, if it were not the *same* (rigid) rod when we apply it to interval Y as it was when we applied it to interval X, we would not have the slightest reason for judging the two intervals to be same length, or different lengths. Now, what sorts of things are (identically) the same from one moment to the next? Answer, 3D things. Using what is now accepted terminology, 3D things *endure*, while 4D things *perdure*. (For further discussion of this distinction, a clear definition of "endurance", and the philosophical desirability of retaining and making use of *both* the concepts of endurance and perdurance, see McCall and Lowe (2006).) Something perdures in virtue of having different temporal parts at different times. It extends temporally. In the same way, for example, a road extends spatially in virtue of having different spatial parts. But how do we know whether two different spatial parts of the same road are of the same length? Only by taking a 3D ruler and transporting it from one region to another. The only criterion for saying two intervals are congruent is to apply a measuring device (rod or clock) to them and see if they agree. And this measuring device can't be a 4D object, it must be a transportable, enduring 3D object. Otherwise you can't measure with it.

To sum up the results of this section, it might be thought that our ordinary 3D notions of physical objects moving, changing and existing through time must give way to more sophisticated 4D conceptions if we are ever to understand the sometimes astonishing effects of special relativity. But in fact the notions of *length* and *duration* rest upon *congruence* and *measurement*, and the latter are inexplicable without 3D measuring devices. In the world of modern science, the 3D and the 4D ontologies are inextricably interlocked, and one cannot be privileged to the exclusion of the other. Instead, both are needed.

4. 3D/4D Equivalence

In this section it is argued that the three- and the four-dimensional ways of looking at the world are even more closely linked. Not only are both needed in science, but they are in fact *equivalent* in the sense that, given a 3D description of something, it can be translated without remainder into a 4D description and vice versa. A 3D description of a tadpole changing into a frog over a period of three weeks can be translated into a 4D description involving an elongated tubular volume with tail-like and leg-like protuberances, immersed in its early stages in 4D water and in its later stages in 4D air. Conversely, the 4D description is translated by taking 3D cross-sections of the 4D volume at

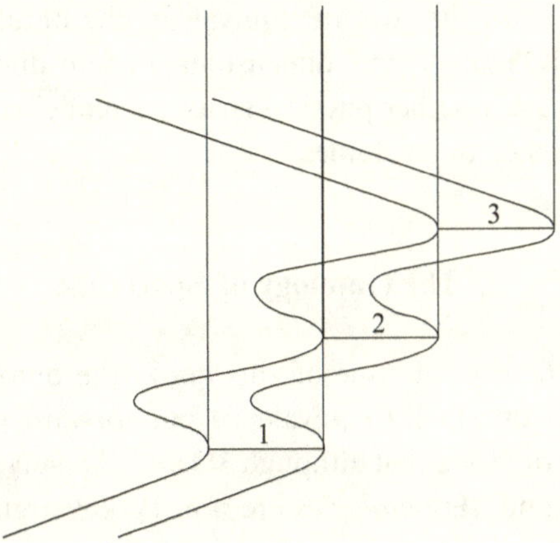

FIGURE 2

different levels, revealing a creature with a tail and gills lower down, and a tail-less air-breathing animal higher up. These 3D sections are "snapshots" of a 3D continuant which endures through time while moving and changing every moment.

Figure 2 is a 4D picture of a yardstick measuring a bolt of cloth. Without the equivalent 3D description, which shows the ruler establishing congruence, no sense can be attached to the assertion that the 4D intervals 1, 2, 3 are of equal length.

True equivalences of the 3D/4D kind, based on intertranslatability, are rare in philosophy. A different example derives from A.N. Whitehead's method of extensive abstraction. In geometry, lines or volumes can be regarded as sets of points, and consequently a philosopher studying the foundations of geometry might come to the conclusion that his ontological primitives were points, and that other geometrical notions could be defined in terms of them. What Whitehead demonstrated, however, was that this definition could be reversed, and that a "point" could be defined as an infinite descending nested set of volumes (Whitehead (1920, chapter 4)). Given the interdefinability of the two notions, any description of a geometrical figure in terms of "points" can be translated into a description in terms of "volumes", and vice versa.

Before Whitehead's discovery, a philosopher who asked whether physical space was "really" made up of points, or "really" made up of tiny volumes, could perhaps be taken as raising a genuine foundational question. But in the light of point/volume equivalence, the ontological issue disappears. In the same way, given the intertranslatability of 3D and 4D descriptions of the world, the ontological question of whether the world is "really" 3D or 4D also disappears. In order to understand physical reality, in particular some of the more bizarre phenomena of the STR, 4D descriptions are necessary. But to understand the concept of 4D length, and its relation to measurement, 3D descriptions are equally necessary. Since the two are equivalent, the ontological question of whether the world is "really" three-dimensional or four-dimensional makes no more sense than to ask whether physical space is "really" composed of points, or "really" composed of tiny volumes.

5. The Ontology of Spacetime

We come now to the central issue of this book: the ontology of spacetime. Consider the thesis of 3D/4D equivalence put forward in the last section. Vesselin Petkov has objected that although 3D and 4D *descriptions* of the world are equivalent, 3D and 4D *ontologies* are not. Petkov argues that since a 3D space and the objects it encloses is a subspace of 4D spacetime, physical reality must be one or the other, but it cannot be both. If we assume that the physical world consists of the totality of presently existing 3D objects, this totality plus the space in which it is embedded constitutes a 3D "simultaneity slice" of 4D spacetime. To assert that this slice constitutes physical reality is inconsistent with relativity theory, for which there is no unique set of events simultaneous with "now". The same goes for objects: the set of objects which exist "now" is observer-dependent, and cannot be taken as constituting the physical world in any objective sense. Petkov's conclusion is that physical reality cannot be three-dimensional, but must consist of the whole of 4D spacetime. There is therefore no "equivalence" between 3D and 4D ontologies.

Petkov's conclusion would be correct if "3D object" meant "instantaneous state of a 3D object". Thus a given observer's simultaneity class is a global 3D space which includes innumerable "objects-at-an-instant", and it might seem that the choice between a 3D and a 4D ontology was the choice between this simultaneity slice and the whole of 4D spacetime. But "3D object" can also mean something quite different, namely a three-dimensional object which endures and exists *through* time as opposed to *at a time*. An object's existing *through* time does not make it into a four-dimensional object. A rabbit which lives for eight years is not a 4D rabbit, but a 3D rabbit which exists continuously over a period of eight years. If the central ontological question is rephrased so as to pose a choice between 4D spacetime, and a universe of 3D objects which exist through time, then the 3D/4D question is put in quite a different light. Since the 3D objects which are proposed as making up the world are not merely objects which exist at *some specified moment*, but instead include all objects which exist *at any time*, relativity theory has nothing to say about whether the ultimate choice of ontologies must be 3D or 4D. There need be no suggestion that 3D existence must be "relativized to an observer", and a 3D ontology can include past individuals such as Julius Caesar, recent ones such as the late Nelson Mandela, and future ones such as the 50th prime minister of Canada.

Concerning the central question which this book addresses, the question of the dimensionality of reality and the ontological status of 4D spacetime, the present chapter takes the position that no choice between 3D and 4D ontologies is forced upon us. That is to say, there is no "fact of the matter" as to whether the world is three-dimensional or four-dimensional. For certain purposes and in certain contexts it is enlightening and revealing to regard it as four-dimensional, as for example when we are trying to understand how a train can be photographed as being both longer and shorter than a tunnel, or how two twins can be different ages when they are reunited after a long journey. For other purposes we are forced to recognize the continued existence and self-identity of three-dimensional objects in time, as when we use a clock to record the length of a journey or a ruler to measure a piece of wood. Not only are the 3D and the 4D descriptions of the world indispensable, they are equivalent, and it is not a question of one being true and the other false. All of which is to say that there is no answer to the question, is the world *really* 3D or 4D? It is *both*, or *either*, depending on the type and degree of understanding that we seek.

6. The Nature of Time

There is one more important topic to be dealt with. According to the thesis of 3D/4D equivalence Jack and Jill can be regarded as 4D objects, or they can be regarded as 3D objects existing through time. But if the latter, what time do they exist in? Since when they reunite Jack is older, it would seem that Jack has lived more than Jill. What if anything does this tell us about the nature of time?

A frequently encountered reaction to the twins paradox is that elapsed time, or proper time, is "path-dependent". Jack and Jill follow different paths in spacetime, Jill's being broken and Jack's straight, and consequently their total elapsed times differ. This is correct as far as it goes, but ignores the fact that *any* inertial path in frame f_2, followed by an inertial path in frame f_3, will be shorter than a single inertial path in frame f_1 if the three paths form a triangle. Imagine for example an infinity of distinct triangles of the same size and shape as triangle ABC of figure 1, located in different parts of spacetime. These triangles are made up of infinitely many line segments A′B′, parallel to and congruent to the inertial line AB in frame f_2, infinitely many segments B′C′ parallel to and congruent to BC in frame f_3, ... etc. Yet they all share the characteristic that in each case the sum of the temporal lengths, or proper times, of the segments A′B′ and B′C′ will be less than the lengths A′C′ by exactly the deficit in Jill's age as compared to Jack's. What this shows is that quantity of elapsed time is common to all congruent segments of inertial lines in a given frame, and therefore that time flow, and elapsed time, are in the most general sense *frame-dependent* rather than *path-dependent* concepts.

What is a spacetime coordinate frame? Ignoring for the moment the metric aspect, i.e. the calibration of the x, y, z and t axes, a frame is essentially a partition of spacetime by parallel hypersurfaces, each of which is orthogonal to the time axis of the frame. Every distinct inertial frame partitions spacetime differently. Our image of spacetime is comparable to the surface of a vast ocean, agitated by parallel wavetrains moving in every direction. The difference between the cosmic ocean and a real ocean is that the cosmic wavetrains do not exhibit interference, but pass harmlessly through each other. At a purely affine level, this is part of the structure of Minkowski spacetime.

When we come to assign a metric to spacetime, a clock is needed to define the relationship of temporal equality of intervals. As discussed earlier, there are no 4D clocks. A 4D volume with regular markings along the time axis provides no guarantee that the markings are equally spaced. There is therefore no substitute for a 3D pendulum clock, or a 3D cesium ion that emits radiation in periodic energy-level transitions, or for that matter a living, breathing person like Jill who keeps time with her heartbeats. Timekeeping is essentially a 3D process, and this fact has implications for the proper understanding of what time is.

There have been in the philosophical literature since the days of McTaggart two very different ways of conceiving time: as an A-series in which different moments possess the changing attributes of "past", "present", and "future", and as a B-series in which moments stand only in the permanent relationship of "earlier" or "later" (McTaggart (1927)). A necessary feature of the A-series is the notion of temporal becoming: a future event *becomes* present, and a present event *becomes* past. Focussing on the A-series creates the impression that time is flowing, although no such impression is created by the B-series. A nice question, connected with but distinct from the problem of the ontology of spacetime, is the question of whether temporal flow is real. This is the question of whether becoming is an objective property of the world, or whether it is a subjective illusion, an idea which conscious beings have but which corresponds to nothing in physical reality. I shall argue that one of the consequences of the twins paradox is that it lends credibility to the thesis that temporal becoming is real, not illusory.

As mentioned above, Jill in her spacecraft, and Jack at home base, act as clocks, keeping time with their heartbeats. At the end of the voyage, it turns out that Jack has experienced many more heartbeats than Jill. In relativistic geometry his 4D inertial path is longer than the sum of Jill's two inertial paths, and the increased length serves as the four-dimensional explanation of why Jack is older: his time-line is longer. But as was stated above, to determine *how much* Jack is older requires a metrization of spacetime, and this in turn requires a clock, human or otherwise. Clocks do two things. First, they measure *temporal length* by (i) establishing the temporal congruence of spacetime intervals, (ii) counting. But secondly, because clocks are 3D objects they undergo a cyclical *temporal process*, the regular periodicity of which qualifies them as measuring instruments. Furthermore, it seems correct to say that it is precisely in virtue of the fact that a clock undergoes this cyclical process that it is able to measure temporal length. If this is so, then, for a clock, *process* is the primary consideration, and measurement of *length* is a by-product, the use to which the process is put. Now, finally, what is process? Here we find the link to temporal becoming.

In the 4D world there are volumes of all shapes and sizes, cut many ways into temporal parts by the partitioning of different frames, but there is no movement and there are no processes. A process is a dynamic thing, and the 4D world is static. Process, like temporal becoming or temporal flow, is an A-concept rather than a B-concept, and is found in the domain of 3D things but not in the 4D world.[2] All of this is to say that if 3D objects evolve by undergoing temporal processes, as they do, and if the cyclical temporal processes of a clock are essential for the definition of temporal length, as they are, then temporal processes must be an objective characteristic of the physical world. They cannot be a subjective illusion, any more than the 4D world can be a subjective illusion. As was said before, both 3D and 4D descriptions are needed if we are to understand physical reality. Finally, a necessary condition of there being temporal processes is that there be temporal flow or becoming. No temporal becoming, no temporal processes. The upshot is that, with care, a complex many-step inference from

(1) 3D clocks measure time
    to
(2) The universe manifests objective time flow
    can be constructed.

A final note. If indeed there is such a thing as temporal becoming, the twins paradox demonstrates that it cannot be universal or global, but must take place within coordinate frames. The temporal passage or "ageing" undergone by Jack in his frame is not the same as the temporal passage undergone by Jill in her frames. Contrary to what Newton says, it is not the case that "Absolute, true, and mathematical time, of itself, and from its own nature, flows equably, without reference to anything external". There is no "universal tide of becoming", in which events across the entire cosmos move in step from "future" to "present" to "past". Instead, each and every inertial frame exhibits its own temporal becoming, and the transition of an event E from "present in frame f_i" to "past in frame f_i" is not the same as the transition of E from "present in frame f_j" to "past in frame f_j". Because of the huge number of frames partitioning space-time, the surge of temporal passage particular to each and every one of these frames is difficult to picture. But it, rather than the absolute or global time

[2] In this chapter I am ignoring the dynamic 4D model of time flow found in McCall (1994). The argument for temporal becoming in the domain of 3D objects of the present chapter is independent of the "argument to the best explanation" found in (1994).

flow which a preferred frame of reference would imply, must be what spacetime undergoes if temporal becoming is a feature of the physical world.

The author wishes to thank the Editor, Dennis Dieks, for valuable discussion and advice.

References

Armstrong, D.M. (1980) "Identity through time", in *Time and Cause*, ed. van Inwagen, 67–78. Boston: Kluwer.

Balashov, Y. (2000) "Persistence and space-time", *The Monist* 83, 321–40.

Bondi, H. (1964) *Relativity and Common Sense*, Garden City, NY: Anchor Books.

Chisholm, R.M. (1976) *Person and Object*, LaSalle, IL: Open Court Pub. Co.

Darwin, C.G. (1957) "The clock paradox in relativity", *Nature* 180, 976.

Geach, P.T. (1965) "Some problems about time", *Proceedings of the British Academy*, reprinted in his (1972) *Logic Matters*, 302–18. Berkeley: University of California Press.

Grünbaum, A. (1963) *Philosophical Problems of Space and Time*, New York: Knopf.

Haslanger, S. (1989) "Endurance and temporary intrinsics", *Analysis* 49, 119–25.

—— (1994) "Humean supervenience and enduring things", *Aust. J. Phil.* 72, 339–59.

Heller, M. (1984) "Temporal parts of four dimensional objects", *Phil. Studies* 46, 323–34.

—— (1990) *The Ontology of Physical Objects*, Cambridge and New York: Cambridge University Press.

Le Poidevin, R. (2000) "Continuants and continuity", *The Monist* 8, 381–98.

Lewis, D. (1983) *Philosophical Papers*, vol. 1, New York: Oxford University Press.

—— (1986) *On the Plurality of Worlds*, New York: Blackwell.

—— (2002) "Tensing the copula", *Mind* 111, 1–13.

Lowe, E.J. (1987) "Lewis on perdurance versus endurance", *Analysis* 47, 152–54.

—— (1998) *The Possibility of Metaphysics*, Oxford and New York: Oxford University Press.

Marder, L. (1971) *Time and the Space-Traveller*, Philadelphia: University of Pennsylvania Press.

Maudlin, T. (1994) *Quantum Non-Locality and Relativity*, Oxford: Blackwell.

McCall, S. (1994) *A Model of the Universe*, Oxford and New York: Oxford University Press.

McCall, S. and Lowe, E.J. (2003) "3D/4D equivalence, the twins paradox and absolute time", *Analysis* 63, 114–123.

—— (forthcoming) "The 3D/4D controversy: A storm in a teacup", *NOUS* 40 (2006), 570–78.

McTaggart, J.M.E. (1927) *The Nature of Existence*, Cambridge: Cambridge University Press.

Mellor, D.H. (1981) *Real Time*, Cambridge: Cambridge University Press.

Perry, J. (1972) "Can the self divide?", *J.Phil.* 69, 463–88.

Quine, W.V.O. (1950/1953) "Identity, ostension, and hypostasis", in *From a Logical Point of View*, Cambridge, MA: Harvard University Press.

—— (1960) *Word and Object*, Cambridge, MA: MIT Press.

—— (1981) *Theories and Things*, Cambridge, MA: Harvard University Press.

Resnick, R. (1968) *Introduction to Special Relativity*, New York: Wiley.

Riemann, B. (1854) "On the hypotheses which lie at the foundations of geometry", reprinted in D.E. Smith (ed) (1959) *A Source Book in Mathematics*, Vol. II, New York: Dover.

Russell, B. (1899) "Sur les axiomes de la géométrie", *Revue de Métaphysique et de Morale* 7, 684–707.

Sider, T. (1997) "Four-dimensionalism", *Phil. Review* 106, 197–231.

—— (2001) *Four-Dimensionalism: An Ontology of Persistence*, New York: Oxford University Press.

Simons, P. (1987) *Parts: A Study in Ontology*, New York: Oxford University Press.

—— (2000) "How to exist at a time when you have no temporal parts", *The Monist* 83, 419–36.

Smart, J.J.C. (1972) "Space-time and individuals", in *Logic and Art: Essays in Honor of Nelson Goodman*, ed. Rudner and Scheffler, 3–20. Indianapolis: Bobbs-Merrill.

Thomson, J.J. (1983) "Parthood and identity across time". *J.Phil.* 80, 201–20.

van Inwagen, P. (1981) "The doctrine of arbitrary undetached parts", *Pacific Phil.Quart.* 62, 123–37.

—— (1990) "Four-dimensional objects", *Nous* 24, 245–55.

Whitehead, A.N. (1920) *The Concept of Nature*, New York: Cambridge University Press.

{ 14 }

The 3D/4D Controversy: A Storm in a Teacup

Storrs McCall and E.J. Lowe

Four-dimensionalists argue that their ontology is philosophically superior to that of three-dimensionalists. For them, the basic ontological elements of the world are 4D "temporal parts" (or 4D "worms", although we shall be concerned primarily with the former). Familiar objects such as trees and paperweights exist through time in virtue of being "mereological sums" or "fusions" of temporal parts. By contrast, three-dimensionalists maintain that the world is made up ultimately of subatomic particles which have no temporal parts, and which persist or endure in such a way that one and the same particle exists at two or more different times.

Although it might appear that philosophers are faced with a clear-cut ontological choice, we shall argue that this is not so. Contrary to what is found in texts such as Sider's *Four-Dimensionalism* (2001), equally precise and rigorous ontological foundations may be constructed using either 3D or 4D building blocks. There are advantages to viewing the world through 4D spectacles in many contexts; for example the train/tunnel and twins paradoxes in relativity theory are more easily comprehensible when explained in 4D terms.[1] But there are also insights to be gained if physical reality is regarded as made up of persisting 3D objects. We argue that the 3D and the 4D descriptions of the world are equivalent in the sense of being intertranslatable without remainder, and take the position that there is no "fact of the matter" as to whether we live in a 3D or 4D world.[2] Instead, one can freely choose whether to describe it in 3D or 4D terms. Either way, it's the same world. The so-called "3D/4D controversy" is a storm in a teacup, and philosophers should not feel obliged to support one side or the other.

---

[1] See McCall and Lowe (2003), and McCall (2004).

[2] McCall and Lowe, *op. cit.*

1. The Definition of "Endurance"

4D objects are extended along the time dimension, and this fact makes it simple to explain the sense in which a 4D object can exist or be located in different regions of time. It does so in the way a 3D thing like a road can be located in different regions of space, namely by having spatial parts. The road from Montreal to Ottawa exists partly in Quebec and partly in Ontario, and similarly the Eiffel Tower, considered as a 4D object, exists partly in the 19th, partly in the 20th, and partly in the 21st century. But a 3D object, which has extension in space but not in time, cannot exist at different times by having temporal parts. The trick is to explain how 3D objects, lacking temporal parts, can persist or endure through time.

The contemporary philosopher who saw the problem of persistence through time most clearly, and who set what have turned out to be the terms of the discussion for the last 20 years, is the late David Lewis. Lewis uses the word *perdure* to describe how 4D objects exist at different times, namely by having temporal parts, and reserves *endure* and *endurance* to refer to the continued existence of 3D objects through time (1986; 202–4). But it is uncertain, following upon Lewis' discussion, that any clear philosophical meaning can be attached to the latter terms. Lewis begins by introducing the neutral word "*persists*" to mean "exists at different times". By his definition, "persistence" applies indifferently to 3D and to 4D objects. But these objects differ in *how* they persist. 4D objects perdure iff they persist by having temporal parts, while 3D objects endure, i.e. persist by being "wholly present" at different times.[3] Perdurance is clear enough, but what does it mean to say that a 3D object is "wholly present" at a time? "Wholly present" normally excludes "partially present", but since it is totally unclear what *this* means, the word "wholly" adds nothing but confusion.[4] Lacking a comprehensible account of how 3D objects persist by enduring, to set beside the perfectly clear account of how 4D objects persist by perduring, we seem to be led to the conclusion that there can be no such thing as 3D endurance through time, and that the notion of endurance is a fiction.

Such a conclusion may please some four-dimensionalists. Sider for example considers it far from clear that three-dimensionalism can be adequately

---

[3] Other philosophers who have used the phrase "wholly present" include Hugh Mellor: "things are wholly present throughout their lifetimes" (1981; 104), and Peter Simons: "At any time at which it exists, a continuant is wholly present" (1987; 175). These and similar references are found in Sider (2001; 63).

[4] In a later article, Lewis glosses "wholly present" in the following way. "[The first solution] lets us say that things persist by *enduring*: the one thing is present at different times; and not mere temporal parts of it, different parts at different times, but *all* of it, wholly present at each of the times." (1988; 65) This adds more confusion, because if anything endures it is a *3D* object, not a 4D object, and as such is not the sort of thing that could meaningfully have temporal parts in any case.

formulated (2001; 63). But this is not so. No doubt, once we accept Lewis' ordering of definitions, from "persist" to "perdure" and "endure", one building on another, the end result is that endurance turns out to be a no-go. But Lewis' ordering is not the only one imaginable. A more rational approach is to drop "persist" as a neutral term and say that to *perdure* is to have temporal parts. 3D objects, not being extended in time, have no temporal parts and consequently cannot perdure. For such an object to *endure*, we shall say, is simply for it to *exist at more than one time*. To this can be added the notion of "continuous endurance". A 3D object X *endures continuously* from t_1 to t_2 (i.e. is a *continuant*) iff X exists at t_1, t_2 and at every time between t_1 and t_2. These definitions are simple yet precise, and rest on no dubious ideas of something being "wholly present".[5]

To avoid any possibility of misunderstanding, it should be clarified what kind of 3D object is in question when we speak of such an object enduring through time. An instantaneous temporal part, for example, is a 3D object, but it is not a 3D object that *endures*, because it exists at only one instant. Properly speaking, an instantaneous temporal part is a 4D object with zero extension along the time axis. It could also be described as a 3D object with a fixed temporal location. But this is certainly not what is meant by a 3D object which endures through time. These points are obvious enough, but should be noted in order to remove any doubts about what kind of thing an enduring 3D object is. We may also speak of such an object as *persisting*, but for clarity it would be best to avoid describing a 4D object as persisting.

Summing up, three-dimensionalism asserts, precisely and unambiguously, that 3D objects persist or endure by existing at two or more different times, not by having temporal parts. In the next section it is shown that enduring 3D objects make up the ontological domain of 3D quantificational semantics. In later sections we argue that a 3D ontology of this kind resolves the problem of change and motion, which Sider deals with using 4D temporal parts.

2. The Domain of Objects in the Formal semantics of First-Order Logic

In many passages in (2001), Sider makes clear the importance he attaches to specifying exactly what sorts of objects first-order quantificational logic adopts as its domain of discourse. He asserts in the introduction that "modern logic's quantificational apparatus mirrors the structure of reality", and that he assumes "an ontology of *things*" (p.xvi). Moreover, he assumes that "there is

---

[5] As phrased, the definition of "endures" applies to 3D physical objects. But it is easily extendible to other types of things, e.g. nation-states, universities, conversations, hockey games, obligations, disagreements, debts, etc.

a single, objective, correct account of what things there are", and emphasizes that "this view of quantification is absolutely central" to his book. The present authors share Sider's respect for quantification theory, but he goes too far in suggesting that quantificational logic, all by itself, implies a thing-ontology over which its variables range, and that it somehow presupposes that there is only one correct account of what sorts of things there are. Considered as a formal system, quantification theory tells us no more than what statements containing quantifiers follow from what. Even when it is provided with a Tarski-style semantics involving a domain of objects, the semantics impose no constraints whatsoever on what kinds of "objects" qualify for domain membership. They could be tables and chairs, or events, or numbers, or linguistic inscriptions. As an ontologist, Sider is free to stipulate that *his* domain is restricted to temporal parts and mereological fusions thereof, but this choice is not dictated by quantification theory or its semantics.

In this chapter we construct a 3D ontology for quantification theory, allowing its variables to range over elementary particles which endure through time but have no temporal parts. Elementary particles combine into atoms and molecules, and atoms and molecules make up macroscopic objects and agglomerations of matter. Since material bodies can retain their identity while continually gaining and losing particles, sets of particles in-a-region, and at-a-time, will be important elements in 3D ontology. Thus Tibbles, the cat, is constituted at any given moment by a set of particles within a spatial volume. Because Tibbles' material constitution varies with time, the set of particles which makes up Tibbles at *t* will generally not be identical with the set which makes up Tibbles at *t'*. We shall speak of Tibbles as being a *sum of particles* at each moment she exists, and we represent by an ordered pair of the form <S, *t*> the momentary sum which constitutes Tibbles at time *t*.[6]

Sums-of-particles-at-a-time need not be considered as new semantic entities, i.e. as new members of the domain, but can be identified with the semantic referents of predicates. As was seen, the basic building blocks of 3D semantics are 3D particles which exist through time. These are the sole elements of the domain. Let S be the set of just those particles which are located within a certain cat-shaped spatial volume at a time *t* and which constitute Tibbles at *t*. Then S is the referent of a monadic predicate F*x* which is read "*x* is one of the particles constituting Tibbles at time *t*". (More technically, S is the set of members of the domain which the assignment function of any model assigns to the predicate F.) A different predicate G, "the moon at time t_2", would be represented by those members of the domain which constitute the moon at time t_2, and so forth. A simple example of a small composite 3D body, which endures

[6] See E.J. Lowe, "Vagueness and Endurance".

through time while gaining and losing particles, would be one constituted at times t_1, t_2, t_3 and t_4 by the following subsets of the domain:

At time t_1: {*a, b, c, d*}
At time t_2: {*a, b, d*}
At time t_3: {*a, d, e, f*}
At time t_4: {*c, d, e, f, g*}.

Each of *a, b, c, d, e, f, g* is an enduring 3D particle.

3. 3D/4D Translatability

The reader will have noticed that there is a close similarity between the set of 3D particles which constitute an enduring object O at a time *t*, and the instantaneous 4D temporal part of O at *t*. This fact provides for a simple translation scheme between the 4D temporal parts ontology and the 3D particle ontology. Let T(O, *t*) be the instantaneous 4D temporal part of O at *t*, and let <O, *t*> be the instantaneous 3D sum of the particles which constitute O at *t*. In 4D ontology, O is the mereological fusion of all its temporal parts T(O, *t*), one for each moment at which O exists. In 3D ontology, O is the set of particles which successively constitute it at each moment O exists, a set which "changes", i.e. is replaced by a new set, each time O gains a new particle or loses an old one. To translate from the 4D to the 3D description of O, reduce O to its temporal parts, and replace each temporal part T(O, *t*) by the momentary sum <O, *t*> of particles which constitute O at *t*. The collection of all such momentary sums <O, *t*>, for every time at which O exists, yields the set of sets of 3D particles which successively constitute O. Conversely, to translate from the 3D to the 4D description of O, first reduce O to the momentary sums of particles which constitute it, then replace each <O, *t*> by the corresponding temporal part T(O, *t*), then reconstruct O as the fusion of its temporal parts.

It may be objected that the 3D/4D translation scheme just described involves cheating at the basic level consisting of the replacement of <O, *t*> by T(O, *t*) and vice versa. It will be said that T(O, *t*) is a 4D object, a 4D object with zero extension along the time axis to be sure, but still a 4D object. What kind of object is <O, *t*>? To say that <O, *t*> is 4D involves cheating, for then the so-called "translation" which replaces <O, *t*> by T(O, *t*) is not a 3D/4D translation but a 4D/4D translation. On the other hand, to say that <O, *t*> is a 3D object seems to contradict section 1, where it was said that a 3D object with a fixed temporal location was not what is meant by a 3D object which endures through time.

This is so. Nevertheless, <O, *t*> can be interpreted in such a way as to restore the integrity of the translation scheme. <O, *t*> may be understood as a 3D object which exists only at time *t* and no other time. It is not an enduring 3D object. It is an instantaneous one, a collection of particles in a spatial configuration which constitutes object O at time *t*. The upshot of this is that the intertranslatability of 3D and 4D descriptions rests ultimately upon entities which can be described indifferently as "instantaneous 4D temporal parts", or "3D objects which exist at one time only". For the 4D ontologist these entities are primitive and basic; for the 3D ontologist they are defined as ordered pairs of sets of enduring particles and times. The one-to-one relationship that exists between them is the foundation of the 3D/4D translation scheme.

4. The Problem of Identity through Change

Sider places great emphasis on the alleged advantages of 4D ontology in coping with problems of identity through time, notably (i) reconciling identity with change, (ii) dealing with temporally coincident objects, and (iii) Hobbes's problem of the Ship of Theseus. In his words (2001; 10):

> If we believe in four-dimensionalism, we can dissolve these and other puzzle cases; if we do not, we are left mired in contradiction and paradox.

No doubt, if Sider could substantiate this claim, philosophers would without hesitation embrace four-dimensionalism. But as we shall show the claim is false. We concentrate on the first and most general problem, that of identity in change, and in what follows give a 3D analysis of change that is as good or better than that of four-dimensionalism.

Aristotle in the *Categories* 4a10–21 remarks that an individual substance can "receive contraries" in the sense of being characterized by mutually contradictory properties at different times. A man can be pale at time t_1 and dark at time t_2. Although Aristotle simply accepts that an individual X which has property A at one time can have property not-A at another time and still be X, other philosophers use Leibniz's Law to deny this. By the indiscernibility of identicals, if Y and Z differ in any of their properties they cannot be identical, and consequently the man who is pale cannot be the same as the man who is dark. The difficulty of attributing contradictory properties such as "pale" and "dark" to a single subject is called by Lewis "the problem of temporary intrisics" (1986; 203–4). His method of resolving it is to deny that there exists an enduring entity with incompatible properties at different times. When Lewis sits at t_1 he has a bent shape; when he stands at t_2 he has a straightened shape. But the thing that is bent is not the same as the thing that is straight. Instead, the thing that is bent is (not Lewis but) a temporal part of Lewis at t_1, and the thing that is straight is a different temporal part of Lewis at t_2. The concept of an enduring 3D entity which changes from being A at one time to being not-A at another time is replaced by a perduring 4D entity with temporal parts. In Sider's words (2001; 5) the four-dimensionalist's "nice solution" to the problem of change defines change as dissimilarity between successive temporal parts.

To this, the three-dimensionalist can reply that there is an equally nice 3D solution to the problem of change. Change is the relative movement, rearrangement, gain or loss of enduring 3D particles in a macroscopic body. In contrast to the 4D account of change as the successive replacement of one unchanging temporal part by another, the 3D account is dynamic. To observe a dynamic change taking place in a 3D object, hold your arm straight, then bend it slowly at the elbow. The particles of the arm move smoothly and continuously from a straight configuration into a bent configuration. The key concept here is "motion". In the 3D world, enduring things move, whereas in the 4D world, motion or more generally change is "replacement of one temporal part by a differing temporal part". Since this chapter argues that one of these two different theories of change and motion is as good as the other, we must examine closely whether a consistent 3D account of change exists.

Lewis, in reply to Lowe (1987), says that he would welcome not one but two tenable solutions to the problem of change: some sort of endurance theory, as well as the perdurance theory (Lewis (1988; 67)). However, despite this conciliatory start, Lewis concludes that in the end Lowe's endurance-based analysis leaves still unresolved the problem of "temporary intrinsics", this time the problem of temporary intrinsic *relations* rather then properties (1988; 69).

When two particles move with respect to each other, they are at one moment separated by one distance, and at another moment by a different distance. When you bend your arm, a particle in your wrist starts off at a distance *d* from a particle in your shoulder, and ends up at a distance *d'* from the same particle. These distance relations, Lewis says, are intrinsic. "The ever-changing distances of particles from one another seem to be temporary intrinsic relations". And, he goes on to ask, "How can the same two things stand in different, incompatible intrinsic relations?" Lewis's answer, of course, is that the things that stand in incompatible relations are not the same, but *different*; they are the different temporal parts of the respective particles at different times. This is the standard perdurantist answer to the problem of varying relative distances between particles.

The 4D answer may be the only tenable one in the absence of a consistent account of endurance. But given the definition of endurance in section 1, it is obvious that an alternative, equally good explanation can be provided of how it is possible for the distance separating two 3D particles to change over a period of time. For the perdurantist, one temporal part of the two-particle system, at t_1, shows the particles separated by distance d_1, and another, at t_2, shows them separated by distance d_2. Equally cogently, the endurantist can now say that the explanation of the change in distance is that the 3D particles are *moving* relative to one another. A prerequisite for motion is *continuous endurance*: two particles that were not continuants over the period t_1 to t_2 could not be said to be in motion during that period. The connection between motion and continuous existence is an important one, and is worth a moment's study. Motion implies something moving. If there were not something that persisted throughout the period of movement, what would it be that moves?

5. A Perdurance-like Surrogate for Motion

Whenever we go to the cinema or turn on the TV, we see something that masquerades as motion, but in fact is the antithesis of motion. This is the rapid replacement of one static image by another. When the stagecoach drives across the screen, the only things moving are the photons emitted by the screen which strike the retina of the eye. The stagecoach image itself does not move, and its apparent motion is an illusion. Rapid replacement of one state by another closely similar state mimics motion, but it is not motion. This also applies to replacement of one instantaneous 4D temporal part by another. What motion requires is a moving 3D object which endures, or so it would seem.

The reasoning in the preceding paragraph may be considered a little too quick, in that the analogy between replacement of images on a two-dimensional cinema screen and replacement of successive temporal parts may not be perfect. For one thing, no 2D entity can be a physical, material object. Therefore, it may be said, a succession of 2D movie or TV images can be at best a poor surrogate of motion. On the other hand, a succession of instantaneous 4D temporal parts might perfectly well represent motion, if motion consists precisely in two objects being a certain distance apart at one time (temporal part A), and a different distance apart at another time (temporal part B).

But is this what motion consists in? Not really. The cinema analogy with temporal part replacement may be questioned because its images are only two-dimensional, whereas instantaneous temporal parts are three-dimensional. A closer analogy is that of an animated hologram. The technology of animated holograms is currently being perfected, but it may not be long before a lifelike 3D laser hologram of a rabbit hopping about on the floor can be created.[7] In this case there is an exact parallel between the successive 3D laser-generated images of the hologram, and the successive 3D temporal parts of a rabbit considered as a four-dimensional volume. When the hologram rabbit twitches his ears, in one image the tips of the ears are 4 inches apart, and in a slightly later image they are 3.9 inches apart. Similarly, when the 4D rabbit twitches his ears, in one temporal part the tips are separated by 4 inches, and in a slightly

---

[7] "... the holographic stereogram process, or multiplex hologram ... enables the artist to take any short sequence of film and create an animated hologram from it". (From "Water Droplet" by Jeffrey Robb, a work in the Jonathan Ross Hologram Collection exhibited at the Butler Institute, January 2000.)

later temporal part they are separated by 3.9 inches. In neither case do the ears move: the only things that move in the hologram are the photons which travel from the hologram to the eye.

The moral to be drawn from the cinema and hologram examples is that images which resemble and succeed one another rapidly can create the illusion of motion. In the same way, 4D temporal parts which succeed one another can create the illusion of motion. But in neither case does anything actually move. The three-dimensional analysis of motion, on the other hand, based on the continuous existence of a moving 3D object which endures throughout the time it is in motion, gives a quite different picture. This is the picture of a physical world of moving objects.

6. Conclusion

To repeat what was said at the beginning of the chapter, we do not advocate that philosophers abandon four-dimensionalism in favour of three-dimensionalism, nor three-dimensionalism in favour of four-dimensionalism. Both ontologies need to be appealed to, if we are to put together an elegant and comprehensive world-view. For some purposes the 4D picture is more illuminating, e.g. in dealing with problems such as the train/tunnel paradox, and for other purposes the 3D picture is preferable. But ultimately it makes no difference which ontological position we adopt. The intertranslatability of 3D and 4D descriptions of the world enables us to move from one ontological stance to the other with ease and confidence. The 3D/4D controversy is indeed a "storm in a teacup".

References

Lewis, David (1986). *On the Plurality of Worlds*, Oxford: Blackwell.

—— (1988). "Rearrangement of Particles: Reply to Lowe", *Analysis 48*, pp. 65–72.

Lowe, E.J. (1987). "Lewis on Perdurance versus Endurance", *Analysis 47*, pp. 152–54.

Lowe, E.J. (2005). "Vagueness and Endurance", *Analysis* 65, pp. 104–12.

McCall, Storrs (2006). "Philosophical Consequences of the Twins Paradox". See Chapter 13 above.

McCall, Storrs and Lowe, E.J. (2003). "3D/4D Equivalence, the Twins Paradox, and Absolute Time", *Analysis 63*, pp. 114–23.

Mellor, Hugh (1981). *Real Time*, Cambridge: Cambridge University Press.

Sider, Theodore (2001). *Four-Dimensionalism*, New York: Oxford University Press.

Simons, Peter (1987). *Parts: A Study in Ontology*, New York: Oxford University Press.

{ 15 }

Laws of Nature and Branching Spacetime

What is a law of nature? An elegant answer to this question is to be found in Armstrong (1983), in which laws are identified with non-logical contingent relations of "nomic necessitation" between universals. For example, the law that all protons have rest-mass *m* takes the form N(F,G), where F and G are the universals "being a proton" and "having rest-mass *m*" respectively, and N is the relation of contingent necessitation that connects them. It is not *logically* necessary that being a proton should imply having rest-mass *m*, but it is contingently and *physically* necessary. Armstrong distinguishes between laws and law-statements, which are attempts to put laws of nature into verbal or mathematical form; on this see also Lehoux (2006).[1] Law-statements can be true or false. When they are true, what makes them true is a law.

In this chapter a different but related analysis of laws is put forward. Laws are based not on universals but on patterns of instantiation of event-types in branching spacetime. The end-result resembles Armstrong's in providing an objective, realist basis for laws of nature, but the branching universe that underlies them is more complex and elaborate than the world of universals. As in David Lewis's theory of Humean supervenience, laws supervene on how events are distributed in space and time. Humean supervenience is "the doctrine that all there is to the world is a vast mosaic of local matters of particular fact, just one little thing and then another" (Lewis (1986), p. ix). In Lewis' picture, laws emerge from the mosaic of events in spacetime, which in Lewis's metaphysics is unbranched. Moving to a more complex branching structure that permits both actual and possible events to belong to a single spacetime history, the account of laws put forward here provides for both (i) the contingency, and (ii) the factual necessity, of laws of nature.

---

[1] Lehoux raises the question of when it became common to describe regularities in the natural world as operating according to strict mathematical laws, and says that this practice dates back only to the 16th and 17th centuries. The matter is discussed in Giere, *Science without Laws* (1999). A mathematical law or equation would be one variety of what Armstrong calls a "law-statement".

Branching spacetime has been described and analyzed in numerous publications by Belnap, McCall, and co-workers: Belnap (1992), Belnap, Perloff and Xu (2001), McCall (1994), McCall (2000), McCall (2009). What branches in branching spacetime are *histories*, i.e. complete four-dimensional differential manifolds that extend back to the beginning of the universe and forward to the end of time (if there is one). The mode of branching between any two 4D histories is along a 3D spacelike hypersurface which serves as the "branch point" between them. For pictures of such branching see McCall (1994), p. 87; (2009), pp. 421–24; and figure 1 below, reproduced from McCabe (2005), p. 666. This diagram is derived by McCabe from Earman (1986), p. 225.

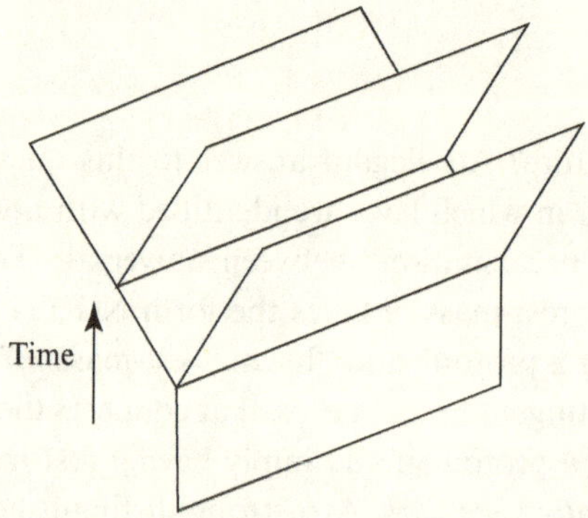

FIGURE 1

The overall structure of the branched universe takes the form of a tree, with the past consisting of a unique 4D trunk, the future of a set of physically possible branching histories, and the present being at the first branch point (figure 2):

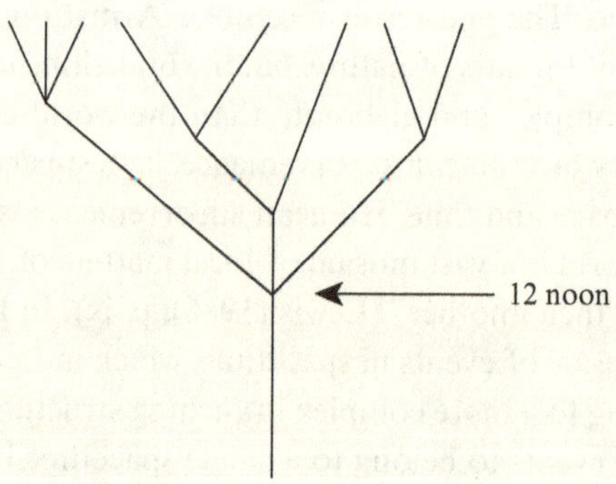

FIGURE 2

As time progresses, a single one of the possible futures at the first branch point gets randomly selected as the *actual* branch, becoming part of the past, while the others vanish. If figure 2 is a model of the universe at time t, figure 3 represents it a moment later:

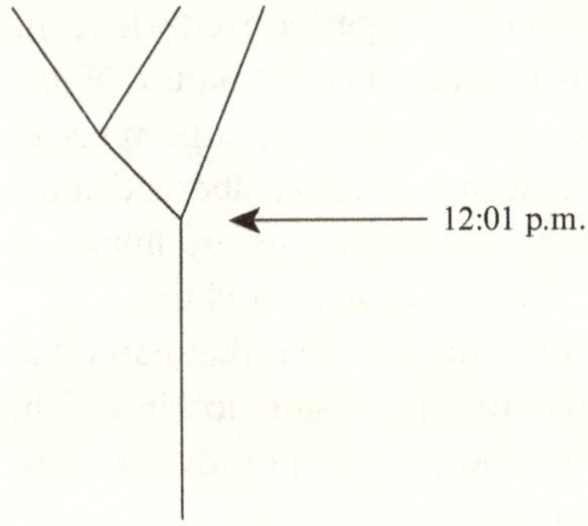

FIGURE 3

This progressive branch loss or attrition represents the flow of time. The model is a dynamic one, in which the first branch point moves steadily up the tree, and sets of possible future histories are successively reduced to a single present and past. The future branches are *real*, but they are not *actual*. There is no danger of running out of branches: for all we know, they extend infinitely far into the future. The universe is a tree that "grows" (or more precisely "shrinks") by losing branches, and progressive branch loss constitutes the flow of time. Far from being a subjective illusion, depending on the existence of conscious beings, time flow is an objective feature of the world. A very different picture of spacetime branching, with no branch loss or attrition, is found in Penrose (1979), p. 592. In this model the branched structure remains fixed and invariant, while the observer moves up the tree.

1. Natural Law Supervenience

Moving now to laws of nature, if throughout the universe-tree at any given period, some set of initial conditions A occurs at a branch point, and if all the branches above A contain B, then during that period B is *physically necessitated* by A:

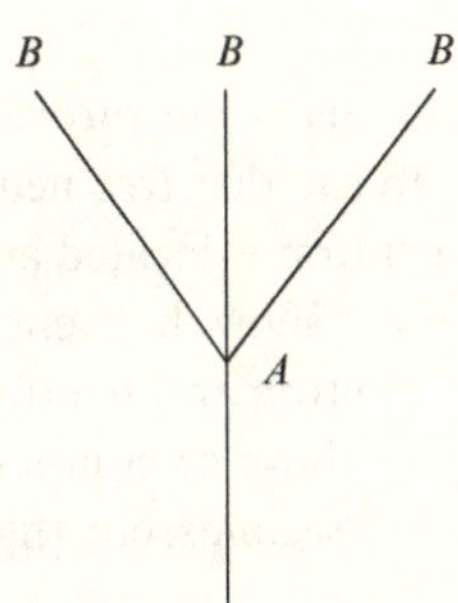

FIGURE 4

If this pattern of A's and B's is repeated everywhere throughout the branching structure, then "All A is B" states a law of nature. What it states is an objective fact about the universe: the law of nature supervenes upon the distribution of events in the world. Armstrong, as stated above, distinguishes sharply between *laws* and *law-statements*. Law-statements are human creations, whereas laws are repetitive patterns of events in space and time.

An interesting feature of this theory is that natural laws can *come into being* with the passage of time. Imagine a situation in which, at the time of the big bang, the pattern of A's and B's was not always as in figure 4, but in some instances followed figure 5:

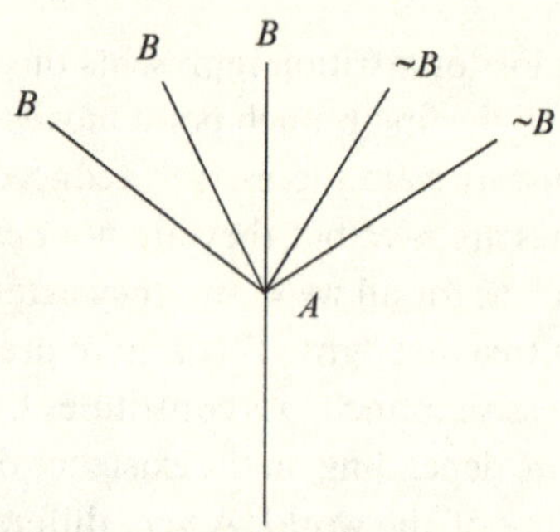

FIGURE 5

Then "All A is B" would not, from the beginning of time, state a law of nature. But suppose further that after a few billion years of branch attrition, no instances of figure 5's pattern remained, leaving only cases in which all branches above A initial conditions were B-branches (figure 4). In that case we could say that "All A is B" had *become* a law. Because of branch attrition, laws of nature constitute a monotonic increasing set. No laws are ever lost, and new ones come into being as the universe passes through different epochs.

2. Probabilistic Laws

Not all laws are exceptionless; many are probabilistic. The laws of particle decay fall into this category. To say that free neutrons have a half-life of 886 seconds means that if a free neutron is located at a first branch point, exactly half the branches above that point show the event of the neutron decaying into a proton, an electron, and an electron anti-neutrino within 886 seconds. Then subsequently, half the branches above each non-decay branch show the neutron decaying within another 886 seconds, etc. (figure 6):

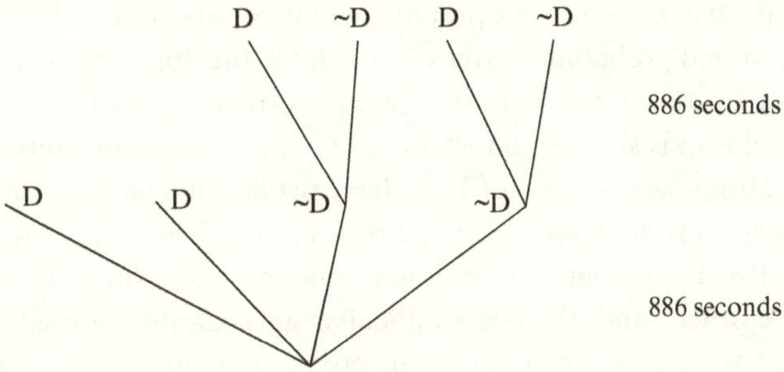

FIGURE 6

A nice feature of the branched theory of laws is that the time interval of 886 seconds, which is an essential component of the law of free neutron decay, is built directly into spacetime structure. This holds also for exceptionless laws. If a copper wire is moved through a magnetic field, an electric current is induced to flow in the wire, not instantly but after a short time lag. The length of the time lag can be found in the branched structure.

3. Non-Rational Probability Values

A different example of a probabilistic law is based on the Stern-Gerlach experiment in quantum mechanics. If a spin-1/2 particle such as an electron in the quantum state |*z*-spin up> passes through a Stern Gerlach magnet slanted at 30° to the vertical *z*-axis, then the electron has a probability of $\sin^2 30° = 3/4$ of exiting in the spin-up channel of the apparatus, and a probability of $\cos^2 30°$ of exiting in a spin-down channel (figure 7):

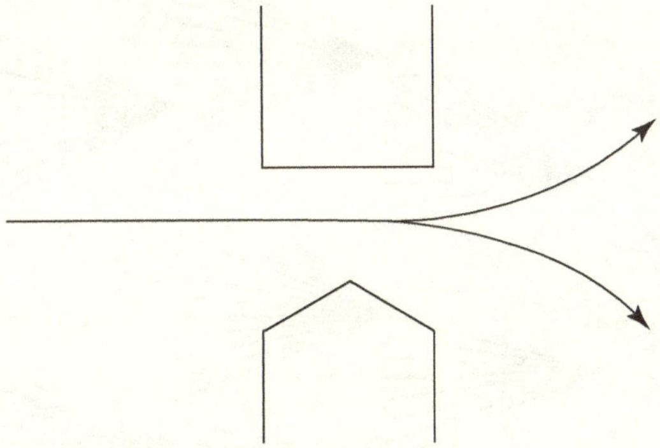

FIGURE 7

In branching spacetime, these probabilities derive from the fact that 3/4 of the branches above the experiment's initial conditions show the electron exiting spin-up, and 1/4 of the branches show it exiting spin-down. In branching spacetime, probability = relative proportionality of sets of branches. This works easily for rational probability values, but what if the apparatus is oriented not at 30° but at 20° to the z-axis? In that case the probability of the electron being measured spin-up is $\sin^2 20°$, an irrational number. How can irrational-valued proportionalities between sets of branches exist in branching spacetime?

This question is unanswerable if we restrict ourselves to finite sets. All proportions between finite sets take rational values only. If infinite sets are considered, on the other hand, there is a difficulty, since Cantor showed in the 19th century that there exist no well-defined proportionalities among infinite sets. One might think, for example, that the proportion of even to odd numbers among the integers was 1/2, but if the integers are ordered differently:

1 2 4 **3** 6 8 **5** 10 12 **7**...

then the proportion of even to odd numbers seems to be 2/3, not 1/2. In Cantorian set theory, there exist no fixed proportions among infinite sets. How is it possible, in branching spacetime, to admit probabilities of future events that take irrational values?

A solution to this problem can be found that does not contradict Cantor's position. If the set of branches above the initial conditions of an electron entering a Stern-Gerlach apparatus possesses a particular kind of structure described as a "decenary tree", then any probability at all of the electron emerging spin-up, whether rational or irrational, can be represented by branch proportionality. A decenary tree is a tree, each branch of which branches in 10 at every one of a denumerably infinite set of branching levels (figure 8, where the time axis extends to the right):

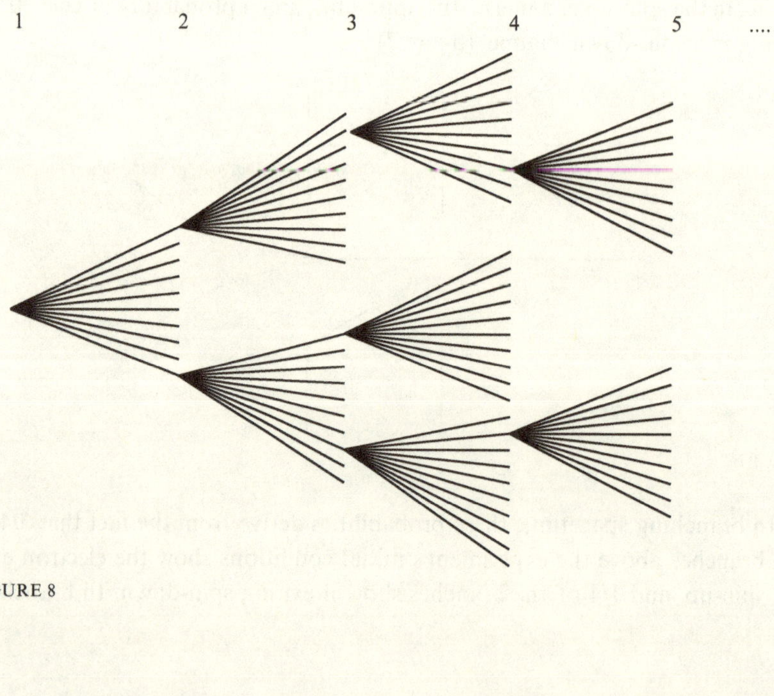

FIGURE 8

A decenary tree contains a non-denumerable infinity of branches, 10 to the power aleph-null.[2] But despite its many branches, a decenary tree need not extend far along the time axis. It may be compressed down into an arbitrarily short but non-zero interval of time, as shown in figure 9:

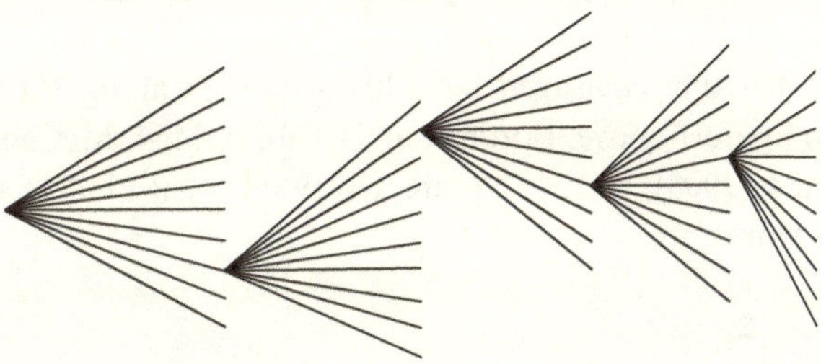

t $\quad t+\frac{1}{2}\Delta t$ $\quad t+\frac{3}{4}\Delta t$ $\quad t+\frac{7}{8}\Delta t$ $\quad \ldots\ldots$ $\quad t+\Delta t$

FIGURE 9

Let the first branching level be at time t, resulting in 10 branches, the second at $t + 1/2\ \Delta t$, yielding 100 branches, the third at $t + 3/4\ \Delta t$, yielding 1000, etc. By $t + \Delta t$ there are infinitely many. A decenary tree provides probability values for outcomes equal to any real number between zero and one, whether rational or irrational, as follows.

An electron with vertical spin has a chance of $\sin^2 20° = 0.884023\ldots$of exiting in the apparatus' spin-up channel when the magnet is tilted at 20° to the vertical, and a probability of $\cos^2 20° = 0.115976\ldots$of exiting in the spin-down channel. Using "u" to denote "spin-up" outcomes and "d" to denote "spin-down", let 8 of the first 10 branches at the base of the decenary tree be u-branches, let 1 branch be a d-branch, and let 1 branch be "open", i.e. be a branch in which the electron is not yet in either channel. Once a branch becomes either u or d, it and all its daughter branches remain u or d for the rest of the decenary tree. At the second branching level, where the "open" branch splits in 10, again let 8 of its descendants be u-branches, 1 a d-branch, and let 1 remain open. At the third level, let 4 of the open branch's descendants be u-branches, let 5 be d-branches, with one left open. Continuing this process, the decimal numbers 0.884023…and 0.115976…are simply reproduced on the branches of the tree, with the result that when time $t + \Delta t$ arrives, exactly $\sin^2 20°$ of its branches are spin-up, and $\cos^2 20°$ are spin-down.[3] The decenary tree

[2] Human beings, who have ten fingers and use decimal number systems, find decenary trees easy to grasp. Binary trees, or ternary trees, would do as well, but are harder to work with.

[3] On top of every one of the branches in a decenary tree, including the "open" branch, there sits at t + delta t the base of another decenary tree. However, the base branch point of that next tree is not "open", but (in the case of the Stern-Gerlach experiment) is either determinately spin-up or determinately spin-down. Although each branch of the lower tree has no "last"

consequently provides all experimental outcomes with precise probability values, whether rational or irrational.

4. Branching along Hyperplanes and Non-Locality

In the spacetime structure envisaged here, histories split along 3D spacelike hyperplanes, as in figure 1 above. Borde et al. (1999), p. 3464, McCabe (2005), p. 673, and Earman (2008), p. 194, picture a very different kind of structure called a "trousers world":

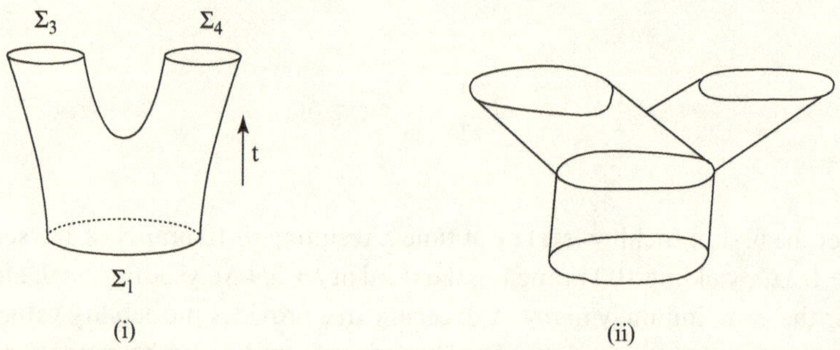

FIGURE 10

In the trousers world (i) there is no well-defined borderline between trunk and branches, whereas in (ii) the branches split along a surface. As will be seen below, splitting along a surface is essential in understanding how branch attrition can throw light on the non-local distant correlations of the EPR experiment. The spacelike hyperplanes along which histories divide are relative to frames of reference. Each one is a constant-time hyperplane in a frame, the events that lie along it being simultaneous with one another in the frame in question. If a model branches along hyperplanes, its branching structure will be different not only at different times (because of branch attrition), but at different *frame-times*. At any given time, the shape of the structure will depend upon the frame of reference or coordinate system used to describe it. The way in which the shape of the universe tree changes from one frame of reference to another will be a "perspectival" change, analogous to the different aspects presented by a 3D object when viewed from different directions. In one frame, the universe model branches along one parallel family of hyperplanes; in another frame, along a different parallel family. These differences can be "transformed away" by changing to a new coordinate system. Time flow, modeled by branch-attrition, will always be time flow *within a frame of reference*.

---

point, every tree at the next level has a "first" point. The universe is, in effect, a huge stack of decenary trees.

This fact provides part of the basis for getting clear about the twins paradox (McCall and Lowe (2003), McCall (2006)).

Proceeding now to the Bell-EPR experiment and the puzzling non-local correlations that accompany it, it will be shown that branching along hyperplanes can elucidate matters. Suppose that a pair of photons, in an entangled quantum state and with anti-correlated polarizations, is emitted in opposite directions from a source S by positronium decay. If the angle of plane polarization of the photons is measured by two similarly aligned polarization analyzers, the two outcomes will always disagree: if the angle of polarization of one photon is found to be "vertical", that of the other will always be "horizontal", and vice versa (Ballentine (1990), p. 448). Now let the analyzers be mis-aligned, say one vertical and the other at an angle φ to the vertical, as in figure 11:

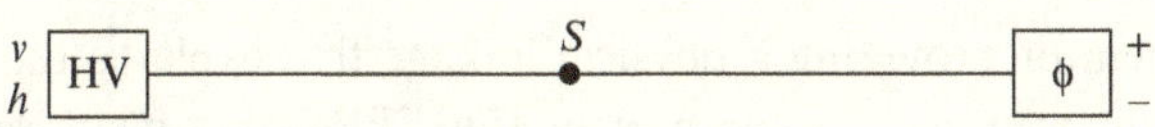

FIGURE 11

The paths of the two photons in branching spacetime are shown in figure 12, in which A is the measurement event of the left photon and B is that of the right photon:

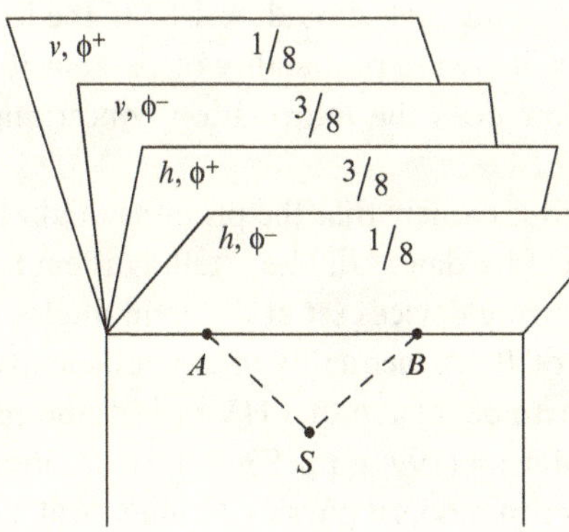

FIGURE 12

Where the possible measurement outcomes are *v* (vertical) and *h* (horizontal) on the left, and φ+ and φ− on the right, the probabilities of the joint outcomes (*v*, φ+), (*v*, φ−), (*h*, φ+) and (*h*, φ−) on left and right are respectively $1/2\sin^2\varphi$, $1/2\cos^2\varphi$, $1/2\cos^2\varphi$ and $1/2\sin^2\varphi$. In figure 12, the figures shown on the four branches represent the *relative proportions* of sets of branches containing the different joint outcomes. When φ = 30°, these values equal 1/8, 3/8, 3/8, and 1/8, and consequently the *probabilities* that the actual branch will be selected randomly from one or another of these sets are also 1/8, 3/8, 3/8 and 1/8.

A particular significance attaches to these four joint probability values: they show that the measurement outcomes on left and right are not probabilistically independent. A necessary and sufficient condition for the outcome *v* on the left to be independent of the outcome φ+ on the right is that the joint probability p(*v*, φ+) should be equal to the product of the two individual probabilities p(*v*) and p(φ+). (If the probability of Smith being left-handed is 1/10, and the probability of Jones being left-handed is also 1/10, then the probability of them both being left-handed, barring some mysterious linkage, should be 1/10 × 1/10 = 1/100.) In figure 12, p(*v*) = p(*v*, φ+) + p(*v*, φ−) = 1/8 + 3/8 =1/2, and p(φ+) = p(*v*, φ+) + p(*h*, φ+) = 1/8 + 3/8 = 1/2. The product of p(*v*) and p(φ+) is 1/2 × 1/2 = 1/4, but the joint probability p(*v*, φ+) equals 1/8, not 1/4, showing that the probability of the left photon being measured *v* is not independent of the right photon being measured φ+. What conceivable mechanism links the two outcomes that could account for the statistical dependence of one upon the other?

The problem of providing a physical linkage that explains the distant correlations of measurement outcomes in the EPR experiment seems impossible, given that the two photons emitted from S are retreating from each other at the speed of light. If the left and right analyzers are both aligned HV, horizontal-vertical, and if the left photon is measured *v*, how does the right photon "know" that it must pass *h*? If the left photon is measured *v*, and if the right analyzer is inclined at 30° to the vertical, how does the right photon know that its probability of passing 30+ should be 1/8? If the left photon had been measured *h* instead of *v*, its twin's probability of passing 30+ would have been 3/8 rather than 1/8. How does the information concerning the left outcome reach it in time?

A possible explanation, namely that the photons might be accompanied by instruction sets, or "local hidden variables", telling them how to be measured if they encounter measuring devices set at different angles, is ruled out by the experimental violation of Bell's inequality in any sufficiently long run of trials, when the left polarizer can be set at either HV or 30°, and the right polarizer at either 30° or 60° (see McCall (1994), pp. 98–99). In addition, faster-than-light signaling is a non-starter in modern physics. Failing local hidden variables and faster-than-light communication, how can the observed distant correlations between the left and right measurement outcomes be accounted for?

There remains one possibility: that the distant correlations are the result of branch-attrition in branching spacetime, where the branches divide along spacelike hyperplanes. In this process a single "actual" future is selected instantaneously along a three-dimensional dividing surface that extends across the entire universe. If both the left and the right EPR measuring events are located on that hyperplane, and if the two polarization analyzers are both aligned HV, then in the branching structure every branch above the hyperplane contains a *v*-outcome on the left if and only if it contains an *h*-outcome on the right. In this way, no matter how far apart the analyzers are positioned, the two outcomes will always disagree.

If on the other hand the two analyzers are mis-aligned, the left one being HV and the right inclined at an angle φ to the vertical, and if (an additional complication) the exact setting of the right one is not made until the photons have already left the source, the branching structure that explains the observed results is that of figure 13:

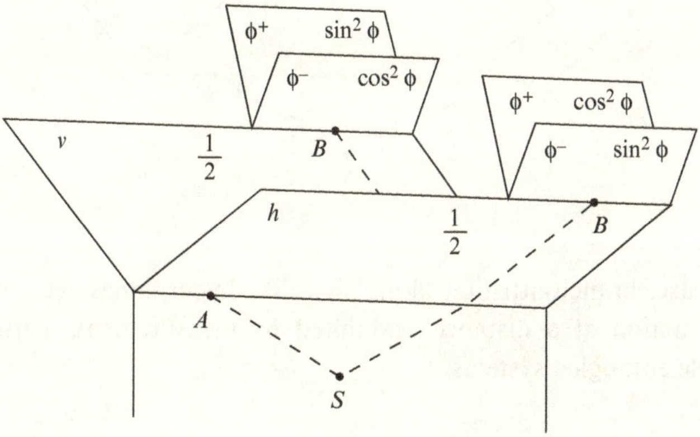

FIGURE 13

Since the photons are moving in opposite directions at the speed of light, the two measurement events are spacelike separated: neither lies within the light-cone of the other. Therefore it is always possible to find a frame of reference *f* in which the left measurement at point A occurs before the right measurement at B, as in figure 13. In frame *f*, since at A no measurement of the right photon has yet taken place, the branches above the constant-time hyperplane that passes through A are equally divided into *v*-branches and *h*-branches. Meanwhile the track of the right photon continues through all these branches until point B, where the right measurement occurs and the structure again divides, this time into φ+ and φ− branches. How does the right photon learn what its probability is of being measured φ+ or φ−?

Well, it has no choice. Since in frame *f* the left measurement has already taken place, because of branch-attrition the right photon finds itself either on an actual *v*-branch (if that was the outcome of the left measurement), or on an actual *h*-branch. If the former, its probability of passing φ+ at B is $\sin^2\varphi$, and if the latter, its probability is $\cos^2\varphi$. These probabilities are built into the model as seen through the perspective of frame *f*. If on the other hand we were to ask how the left photon knew what *its* probability of passing either *v* or *h* was, given that the right measurement had already taken place, the answer would be found in the perspectival branching structure based on a frame in which B occurs before A (figure 14):

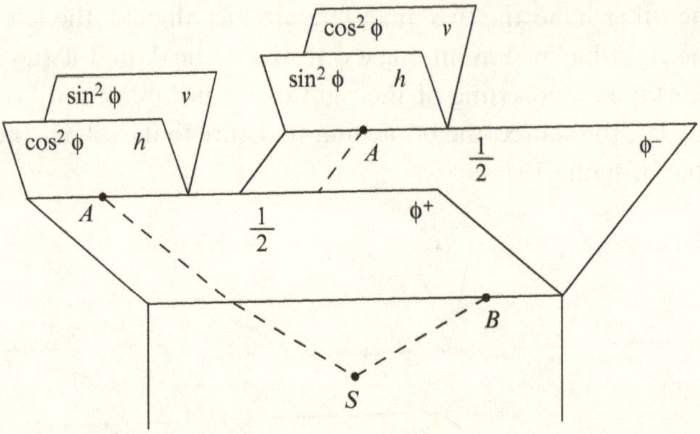

FIGURE 14

In either case, branch-attrition along spacelike hyperplanes accounts for the apparent "action at a distance" exhibited by measurements performed on two-particle entangled systems.

5. Branching Spacetime and the Hausdorff Property

A topological space is called a Hausdorff space if for each pair *x, y* of distinct points of the space, there exist disjoint neighborhoods N*x* and N*y* of *x* and *y*. If two spacetime histories H_1 and H_2 branch, and if their mode of branching is what is known as "lower cut", then their union is a Hausdorff space. But if the mode of branching is "upper cut", their union is not Hausdorff. The names lower and upper cut were originally used by Belnap, and the difference between them is as follows (McCall (1994), pp. 289 ff.). In a Y-shaped space, the division between the two one-dimensional arms of the "Y" is *lower cut* if the least upper bound of the trunk of the "Y" coincides with the greatest lower bound of each of the branches. But if the trunk has no least upper bound, i.e. is an open set of points, while each of the branches possesses its own distinct greatest lower bound, then the branching is *upper cut*. See figure 15, where (iii) is from Visser (1996), p. 252, and McCabe (2005), p. 668:

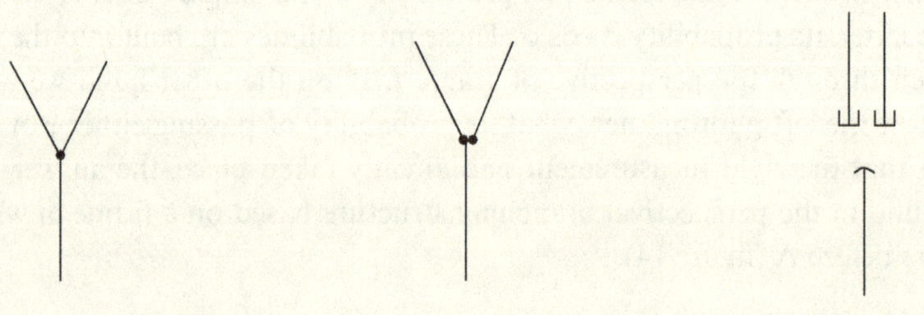

(i) Lower cut (ii) Upper cut (iii) Upper cut

FIGURE 15

As figure 15 shows, lower cut branching is Hausdorff and upper cut is non-Hausdorff. Around any two distinct points x and y belonging to (i), there exist disjoint open sets Nx and Ny that do not overlap. This is not true of the two points x and y in (ii) and (iii) that are the greatest lower bounds of the left branch and the right branch respectively. Any neighborhood of x includes points belonging to the trunk, and the same is true of any neighborhood of y. So any such pair of neighborhoods overlap, and the two points x and y are not Hausdorff separated.

As far as real, physical spacetime is concerned, e.g. as dealt with in general relativity, how seriously should a failure of the Hausdorff property be taken? Very seriously, in the opinion of many experts. We find in Earman (2008), pp. 199–200, the following:

> The assumption of Hausdorffness is explicitly invoked only sporadically in textbooks on general relativity. But it is implicitly assumed in so many standard results in GTR that dropping it would require a major rewriting of textbooks. Here are two examples of widely used results that depend on Hausdorffness. (i) A compact set of a topological space is closed—if the space is Hausdorff. (ii) If a sequence of points of a topological space converges the limit point is unique—if the space is Hausdorff. The situation is best summed up by a dictum of Robert Wald… "Asking what relativistic physics would be like without Hausdorffness is like asking what the earth would be like without its atmosphere".

Given the importance of the Hausdorff property, it would seem either that the idea of spacetime branching should be abandoned entirely, or, failing that, that only lower-cut branching should be recognized. Earman would favour the first alternative. But in connection with the second, an equally important consideration is that lower-cut branching spaces are not locally Euclidean. A space is "locally Euclidean" if it is homeomorphic to a Euclidean space for all sufficiently small neighborhoods around any point. Around a lower-cut branch point, however, a space cannot be Euclidean for any neighborhood no matter how small (see figure 15). In a Y-shape, no neighborhood around the branch point can be part of a one-dimensional Euclidean line, and a lower-cut branching space is not locally Euclidean. McCabe lists what must be given up if one works with a branching non-locally-Euclidean space:

> There is no tangent vector space at any of the branch points; one cannot take the sum of the vectors which are tangent to different branches emanating from the same point. There is therefore no tensor algebra at these points either. The Einstein field equations, the energy conditions of general relativity, and the expression for the local conservation of energy-momentum, cannot hold at the branching points because these expressions and equations are tensorial. (McCabe (2005), p. 670)

Supporters of branching spacetime are confronted with a dilemma. If the branching is upper-cut, the Hausdorff property is violated, whereas if it is lower-cut, the space is not locally Euclidean. Which to choose? I would suggest choosing upper-cut branching, for the reason that every separate complete history in upper-cut space is a differentiable, locally Euclidean, Hausdorff manifold. Physics can be done on such manifolds. By contrast, individual histories in lower-cut branching spacetime are not locally Euclidean at branch points, and at these points the Einstein field equations do not hold. In opting for upper-cut spaces that are everywhere locally Euclidean, Hausdorffness is sacrificed only at the level of the entire branching space. Each individual history, to which the Einstein equations apply, is Hausdorff.

6. Summing Up: The Pros and Cons of Invoking Branching

The time has come to weigh the advantages of spacetime branching against the disadvantages, and to reach a conclusion. It has been argued that a dynamic branching structure, with branch attrition, can provide an ontological basis for laws of nature; i.e. what makes true law-statements true. Such a structure exhibits a graphic difference between past and future, and provides an objective model of time flow. It allows for the possibility that natural laws may evolve through time, with new laws coming into being in the future. In addition, the particular way in which future histories branch in decenary trees creates numerical values for relative proportionalities of sets of branches, and provides every future event, and every probabilistic law, with an exact, non-subjective probability value. To the author's knowledge, nowhere else can we find an example of a precise numerical value being encoded ontologically, in the physical world. The theoretical values of the probabilities of outcomes of quantum experiments may be calculated using vectors in Hilbert space. But what is it about the real world that makes experimental outcomes conform to their theoretical Hilbert-space probabilities? Where is the real—as opposed to the theoretical—probability value to be found? I suggest that it can be found in real proportionalities of real sets of future spacetime branches.

Finally, branching along hyperplanes can explain the long-distance correlations between outcomes of measurements performed on pairs of particles in entangled quantum states. Branch attrition along such hyperplanes furnishes what amounts to instantaneous communication of information across arbitrarily large distances, without anything "travelling" at superluminal speeds. If spacetime were not to branch, none of these things would be possible. The conclusion of the chapter is strongly to suggest the incorporation of branching into the vocabulary of theoretical physics.

References

Armstrong, D. (1983), *What is a Law of Nature?*, New York: Cambridge University Press.

Ballentine, L. (1990), *Quantum Mechanics*, Englewood Cliffs, NJ: Prentice-Hall.

Belnap, N. (1992), "Branching space-time", *Synthese* 92, pp. 385–434.

Belnap, N., Perloff, M. and Ming Xu (2001), *Facing the Future*, New York: Oxford University Press.

Borde, A. et al. (1999), "Causal continuity in degenerate spacetimes", *Classical and Quantum Gravity* 16, pp. 3457–3481.

Earman, J. (1986), *A Primer on Determinism*, Boston: Kluwer Academic.

—— (2008), "Pruning some branches from 'Branching Spacetimes'", in D. Dieks (ed.) *The Ontology of Spacetime* II, pp. 187–205. Boston: Elsevier Science.

Giere, R.N. (1999), *Science without Laws*, Chicago: University of Chicago Press.

Lehoux, D. (2006), "Laws of nature and natural laws", *Studies in History and Philosophy of Science* 37, pp. 527–49.

Lewis, D. (1986), *Philosophical Papers*, vol. II, New York: Oxford University Press.

McCabe, G. (2005), "The topology of branching universes", *Foundations of Physics Letters* 18, pp. 665–76.

McCall, S. (1994), *A Model of the Universe*, New York: Oxford University Press.

—— (2000), "QM and STR: The combining of quantum mechanics and relativity theory", *Philosophy of Science* 67 (Proceedings), pp. S535–S548.

—— (2006), "Philosophical consequences of the twins paradox", in D. Dieks, (ed.) *The Ontology of Spacetime*, pp. 191–204. Boston: Elsevier Science. See Chapter 13 above.

—— (2009), "Objective quantum probabilities", in Greenberger, Hentschel and Weinert (eds) *Compendium of Quantum Physics*, pp. 420–425. New York: Springer.

McCall, S. and Lowe, E.J. (2003), "3D/4D equivalence, the twins paradox and absolute time", *Analysis* 63, pp. 114–123. See Chapter 12 above.

Penrose, R. (1979), "Singularities and time-asymmetry", in S.W. Hawking and W. Israel (eds) *General Relativity: An Einstein Centenary Survey*, pp. 581–638. Cambridge and New York: Cambridge University Press.

Visser, M. (1995), *Lorentzian Wormholes: From Einstein to Hawking*, Woodbury, NY: American Institute of Physics.

{ 16 }

Objective Time Flow

Storrs McCall

1. The Mind-Dependence Theory of Temporal Becoming

Is there such a thing as temporal passage, which is independent of the existence of conscious beings? Or is time flow "subjective" in being dependent upon conscious perception in the way John Locke maintained colors were? In this section, Adolf Grünbaum's mind-dependence theory of temporal becoming[1] will be presented, together with his criticisms of Reichenbach, who takes an objectivist position on time flow. In the next section an alternative to the mind-dependence theory will be put forward.

The classic statements of the mind-dependence theory are those of Eddington[2] and Weyl:

> Events do not happen; they are just there, and we come across them. "The formality of taking place" is merely the indication that the observer has on his voyage of exploration passed into the absolute future of the event in question. ([6], p. 51)
>
> The objective world simply *is*, it does not *happen.* Only to the gaze of my consciousness, crawling upward along the life line of my body, does a section of this world come to life as a fleeting image in space which continuously changes in time. ([25], p. 116).

An earlier version of this chapter was read at the annual meeting of the Canadian Philosophical Association in 1972. The author has benefitted greatly from discussion with Adolf Grünbaum, Nicholas Rescher, Nuel Belnap, Hector Castaneda, Rich Thomason, Jerry Massey, Paul Fitzgerald, Ferrel Christensen, Allen Janis, Jeffrey Winicour, and Don Maier.

[1] The most recent statement on this theory is [10], in which references to earlier versions in [11] and [8] will be found.

[2] Despite his clear statement of it, Eddington did not himself subscribe to the mind-dependence theory, believing it to hold true only of a deterministic metaphysics.

The universe, on this theory, is as it is represented on a Minkowski diagram; a four-dimensional spacetime manifold in which are located all the events and all the things that have been, are, or ever will be. Of course, nothing on the Minkowski diagram corresponds to "have been", "are", or "will be". Events simply occur in spacetime in the sense of having spatio-temporal locations.

Grünbaum asks the question, what is it, over and above its occurrence at a clock time *t*, that characterizes a physical event *E* as *now* or as belonging to the present? His answer is that what is necessary so to characterize *E* is that at *t* some mind-possessing organism *M* is conceptually aware of experiencing either *E* or some other event simultaneous with it in *M*'s reference frame ([10], pp. 206–7). The characterization is mind-dependent: if there were no conscious beings, there would be no present and no "now". Furthermore, the observer's "now" line divides Minkowski spacetime into two submanifolds, and which of these should be called "past" and which "future" is also a mind-dependent matter. If there exist irreversible or other time-asymmetric processes in nature, then time will be anisotropic, and the observer, looking at the universe, will see a structural difference or "grain" along the time axis. But he will see no label indicating "this end up." The selection of future and past, like the designation of "now", is for Grünbaum a mind-dependent choice, not dictated by anisotropy.

In opposition to the mind-dependence theory, Reichenbach maintains that if there is such a thing as temporal passage at all, it must be a phenomenon which is of physical and not merely psychological significance. For Reichenbach, that which bestows physical significance upon temporal becoming is the indeterminism of modern physics. The past, which is the region of events that have already taken place and that are therefore specific and identifiable, stands in contrast to the future, where in place of specific events, we have only probabilities of events. The present, which divides future from past, is the point at which probabilities change into actualities:

> The concept of "becoming" acquires significance in physics: the present, which separates the future from the past, is the moment at which that which was undetermined becomes determined, and "becoming" has the same meaning as "becoming determined".[3]

Reichenbach's theory of objective becoming, though at first sight attractive, contains a fatal weakness. As Grünbaum points out, what has been overlooked is that being determined is a *relative* matter. There is no such thing as being determined *simpliciter*; an event is determined or undetermined only relative to another set of events. Once this is recognized, Reichenbach's theory fails to single out any unique present, which constitutes a metaphysical divide between past and future. Instead, it singles out many such "presents". There is

[3] [18], quoted in [11], p. 320. See also [19]. The sense in which Reichenbach is using the word "determined" is discussed below.

the present of 1800, which divides the past-of-1800 from the future-of-1800, the present of 1976, and the present of 2000. If Reichenbach were to protest that the "presents" of 1800 and 2000 were different in that the indeterminacies of 1800 had already been translated into determinacy, while the determinacies of 2000 did not yet exist, we would ask, with Griinbaum, for whom? The reply can only be, for ourselves, now. Hence Reichenbach's criterion fails to distinguish the presents of 1800 and 2000 from the present of 1976 unless we appeal to a quite different criterion of the present, namely the subjective criterion of conscious experience. All of which brings us back to the mind-dependence definition of *now*-ness.

Reichenbach fails, then, to erect any objective difference between past and future. In particular, it will not be possible for him to say that events future to an observer *O* which are *undetermined* with respect to him are also *indeterminate,* or lack what Grünbaum calls *attribute-specificity.* For event *E* to be determined with respect to *O*'s present means that there exists a certain type of functional connection linking the attributes of the events of *O*'s present to the attributes of *E* (cf. [10], p. 223). Plainly lack of such a functional connection does not imply that *E*'s attributes lack specificity. In spite of this, Reichenbach's use of the word "determined" seems to indicate that he tacitly identifies "determined" with "determinate". For certain past events might be entirely undetermined vis à vis present events, in the sense of having no functional connection with them, in which case it would not be true to say, in the orthodox sense, that the past was "determined." What then can Reichenbach mean? The most reasonable hypothesis is that when he speaks of the present or the past as "determined", he means that all its events are (now) determinate or attribute-specific, whereas at least some future events are not. "Becoming" would then mean "becoming attribute-specific". If this is so, then Grünbaum is entirely right to protest ([10], p. 223) that the question of whether the universe is deterministic or indeterministic is *not* the question of whether some or any events lack attribute-specificity. Indeterminism can be true even though all events are equally determinate or attribute-specific, hence indeterminism alone cannot yield the ontological difference between past and future that Reichenbach seeks. What more is needed will emerge in section two.

2. An Objective Model of Temporal Becoming

If time flow is not to be dependent upon the existence of conscious observers in the universe, a theory or model of it must be devised which is ontological rather than epistemological. The theory presented here is based upon the notion of a "universe-picture", namely a complete state-description of the world including past, present and future, and is "dynamic" rather than "static" in a sense to be defined below. The dynamic element of the theory serves as a representation of time flow. At no point does the theory make any appeal to mental phenomena or conscious experience.

We shall begin by presenting not one theory but four, although only the last of these will be dynamic and hence capable of modeling time flow. The others serve as introductions to it. We start by giving a diagrammatic representation of the four theories, since the salient features which distinguish one from another can be represented pictorially.

Theory *A*, the Minkowskian view, is the one that underlay much of the discussion of the preceding section. According to it the universe is a four-dimensional spatio-temporal manifold in which are located all past, present and future events. In the diagrammatic representation

(*A*) Traditional Minkowskian View

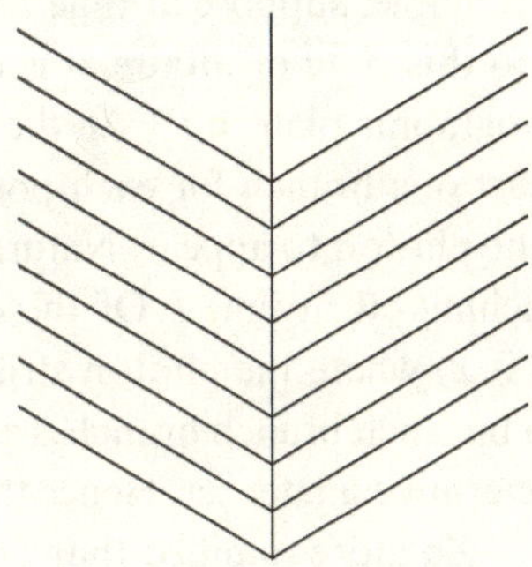

(*B*) Distinguished Branch Theory

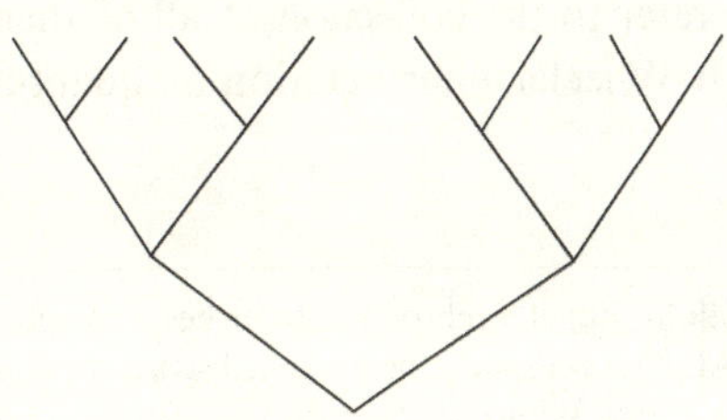

(*C*) Multiple Reality Theory

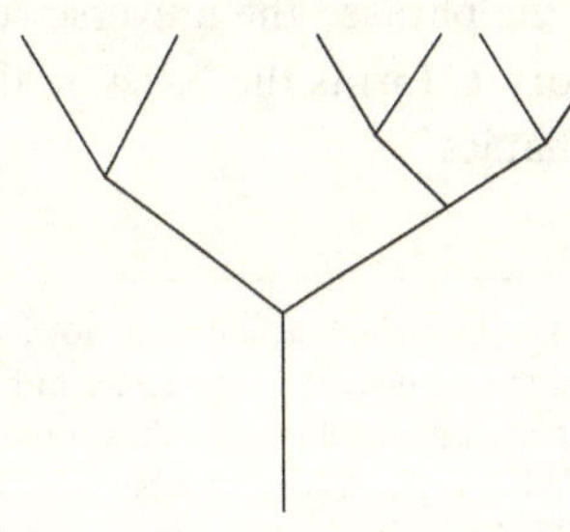

(*D*) Dynamic Theory

of it as a straight line, the time axis is to be taken as extending vertically up the page. Each event in the Minkowski world is completely "attribute-specific", and whatever knowledge human beings may or may not have of these events is irrelevant to their occurrence in the manifold. The Minkowskian view is indifferent as between determinism and indeterminism. If the right sorts of functional connections link event-attributes the Minkowski world is deterministic; if not, not.

Theory *B*, the distinguished branch theory, is like all the remaining theories an indeterministic model. It represents the universe as a branched structure, each of whose branches is a four-dimensional Minkowski manifold.[4] One of these branches (the main stem in the diagram) is "distinguished" in the sense that it represents the actual history of the universe: it is precisely the world of Theory *A*. The other branches represent "histories" or courses of events which the world might have followed but did not. The set of branches that fork off from the main stem at each moment in time consists of the set of world histories which are physically possible relative to the state of the world at that instant.[5] For example, suppose at time t a photon is passing through a slit. Since, relative to this state of affairs, it is not determined where the photon will strike a photographic plate at $t + \Delta t$, the set of possible futures relative to t will include at least one branch for each point on the plate where it is physically possible for the photon to appear. Naturally there will be a great many possible futures branching off at time t. Of these, the main branch represents the actual point (call it x) where the photon strikes the plate. Furthermore, not only the main stem but each branch branches at every point where an earlier set of events fails to determine a later set. Hence the branching of the universe-picture of Theory *B* is much more ramified than appears on the diagram.[6] But one and only one of these branches is distinguished as the "actual" branch.

It is this last feature of Theory *B* that is lacking in Theory *C*, the "multiple reality" theory. Here there is exactly the same branching as before, but no branch is distinguished as the actual one. Instead, all branches are equally real, and the phrase "the universe" can only refer to the collection of all of them. Theory *C* forms the basis of the Everett-Wheeler interpretation of quantum mechanics.[7]

---

[4] The branched structure is not itself a manifold, but if each of the branches had been a four-dimensional Euclidean space rather than a Minkowski space, the branched structure *would* have been a manifold. I owe these observations to Jeffrey Winicour.

[5] What is physically possible is to be distinguished from what is logically possible. (See [14], pp. 427–429.) Given that I am now sitting in Pittsburgh, it is logically but not physically possible that I should be in Washington one second from now. The set of branches which are logically but not physically possible relative to a given world-instant plays no role in any of the theories considered here.

[6] If each branch consists of a continuum of world-instants, and if each such instant is a branch-point, then the number of branches is of cardinality higher than that of the real numbers. See [15].

[7] See [4], which contains the principal expository article [3] as well as almost everything else so far written on the Everett-Wheeler interpretation. I am indebted to Nicholas Rescher, who discusses the Everett-Wheeler theory in Chapter 5 of [20], for originally drawing my attention to it. Rescher refers to the universe in question as the "Borges world", commemorating Jorge Luis Borges's ideas as portrayed in his story "The Garden of Forking Paths". Borges envisages a fictional narrative in which all alternative possibilities are concurrently realized.

Theories *A, B* and *C* differ markedly from each other, but they all have something in common. This is, that the universe-pictures associated with them are static or time-invariant in the sense that the number as well as the arrangement of branches is a permanent structural feature of them. Theory *D*, on the other hand, differs from the others in that the mode of branching of the universe is not a permanent structural feature of it, but varies with time. We thus have not one but many universe-pictures, each one being a complete state-description of the universe *at a time*. Every picture has the form of a tree, the spacetime manifold which forms the trunk containing all past and present events (relative to the time in question), and the branches representing all physically possible courses of future events. Like Theory *C*, and unlike Theory *B*, no one of these branches is distinguished or "preferred" vis-à-vis its neighbors. All branches are equally real, and nothing indicates which branch will eventually form part of the trunk in a later universe-picture. The universe (at each time) is the whole tree; trunk and branches together. But unlike Theory *C*, Theory *D* does not allow reality to be multiple. There do not, according to Theory *D*, exist many different "worlds" (three-dimensional cross-sections of four-dimensional manifolds), some differing only minimally from our world, some a lot, in each of which the time is (say) 11 a.m., April 4th, 1976.

The dynamic feature of Theory *D*, which differentiates it from the other theories, consists in the following. The complete state-description of the universe, i.e., the universe-tree, is different at different times. The difference consists in this. If t_2 is later than t_1, the universe-tree of t_2 is a proper subtree of the universe-tree at t_1. To take a simplified example, the universe in the years 1800, 1900, and 1976 might have the structures pictured below. Each tree is a subtree of the tree(s) to the left of it, as is plain upon redrawing the last tree:

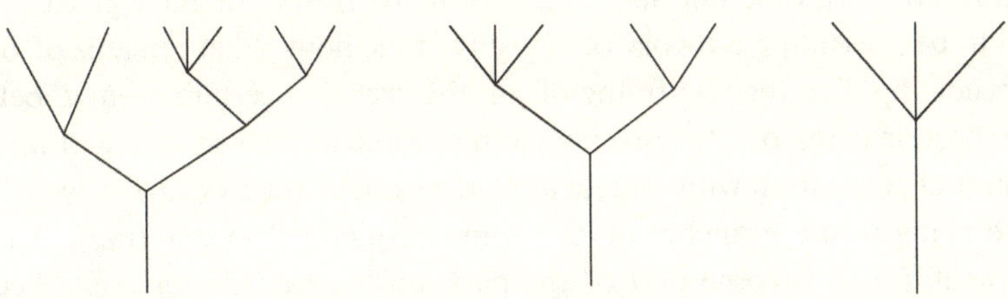

1800 1900 1976

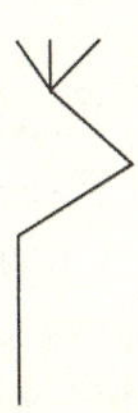

1976

Each moment of time, therefore, defines a separation of the universe into past and future that is ontological rather than epistemological. Furthermore, each moment in time defines a *different* separation. It is this feature of Theory *D* that models time flow.

3. Relativistic Time-Flow

Theory *D* is as yet incomplete. The feature that distinguishes it from Theories *A*–*C* is that it allows no complete description of the universe to be time-invariant. Instead, each universe-picture is a picture of the universe at a certain time. Although, generally speaking, each model will have a branching future and a non-branching past, we do not yet know how to decide whether an arbitrary event belongs to the "future" or the "past". According to the theory of special relativity, there exists no absolute, global way of dividing events into past, present and future. Instead, (*i*) relative to every spacetime point the double light-cone at that point divides events into "past", "future", and "elsewhere", and (*ii*) relative to each instant of time in each coordinate frame, the space-like hypersurface orthogonal to the time axis at that instant divides events into "past", "future", and "present". Of these modes of division, only the second is appropriate here. Since each universe-picture of Theory *D* is a picture of the universe at a time, and since time itself is only specifiable within a coordinate frame, we shall want each universe-picture to be a picture of the universe at a certain frame-time. Past and present events constitute the trunk of the universe-tree, while different possible courses of future events occupy different branches.

It might be objected that, in making each universe-picture of Theory *D* relative not only to a time but also to a coordinate frame, we have given up any claim to be presenting a theory of objective time flow. What remains of objective becoming if in the rest-frame of an observer *O* a certain event *E* belongs to the past or trunk of *O*'s universe-picture, while in the rest-frame of another observer O^*, in motion with respect to *O*, the same event *E* occurs only on some of the many future branches of O^*'s universe-picture? In one frame it might seem as if *E* had become part of the past, or "happened", or even "become real", while in the other frame it had not. Moreover, *O* might accelerate and in so doing change from a rest-frame f_1, in which *E* had already "happened", to a frame f_2 in which *E* had not.

These difficulties, and their implications for temporal becoming, have been discussed by Putnam [17] and Fitzgerald [7]. They argue that since an event *E*'s happening, or becoming determinate or real, is dependent either upon one's spatio-temporal position or upon one's frame of reference, special relativity theory is incompatible with any theory which attempts to ascribe time flow or becoming to the universe as a whole. As Fitzgerald picturesquely puts it:

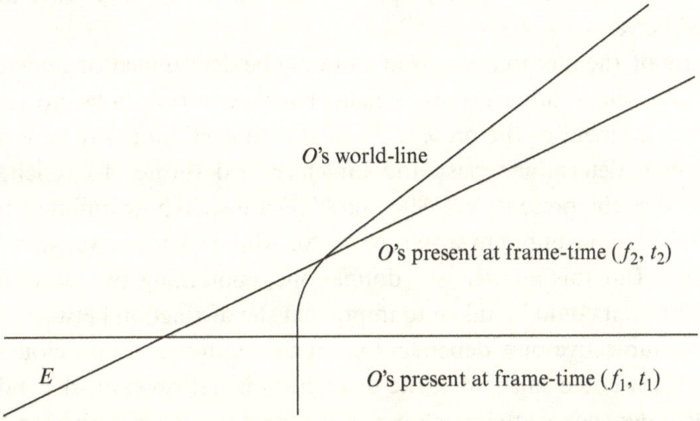

> We no longer have a single tide of "Absolute Becoming" surging into "the future", but an infinite plethora of little crisscrossing ripples, one for each (coordinate)-system. ([7], p. 319)

But it is far from clear that a "single tide" of becoming or time flow is something that any philosopher could reasonably want. Nor is it clear that if there were such a thing as objective time flow, it would manifest itself to observers in a frame-invariant way, any more than spatial or temporal length manifests itself in a frame-invariant way. If temporal passage is to be interpreted as a "becoming", or "coming into being", or "happening" of events at all, then the type of becoming needed will not be frame-invariant. To see this, suppose that the "happening" of each event were represented by the switching on of a tiny light on the Minkowski diagram. Plainly, no random or incoherent pattern of illumination could represent time flow. The only acceptable representation would be an orderly progression of "happenings" up the Minkowski diagram: a regimented march of becoming. But in order for such a march to be regimented, which requires that events be placed in simultaneity classes, a frame of reference is needed. No frame-invariant process of becoming could possibly represent time flow. Therefore it is natural that the universe-pictures of Theory *D* should be dependent not only on a time but also on a frame, and that for some observers certain events should have happened which for others merely lie on possible future branches. This does not make the universe observer-dependent, but it does make it frame-time-dependent.

We are now in a position to answer the question, in what sense is the theory of time flow presented in this chapter an objective one? It will be remembered that Reichenbach's theory of becoming, discussed in Section 1, failed because

Reichenbach was unable to produce any objective difference between past and future. Let us therefore re-examine the arguments against Reichenbach's theory, to see whether or not they apply to the theory of relativistic time flow presented here.

Because of the fact that an event can only be determined or undetermined *relative* to a set of other events, Reichenbach's criterion failed to single out any unique moment as the present. Instead it singled out many presents, each with its own determined past and undetermined future. To Reichenbach's objection that the presents of 1800 and 2000 are somehow different from the present of 1976, Grünbaum would reply, for whom? The answer was, for ourselves, now. But this answer is a double one, containing two quite different answers. First, it could be taken to imply that the distinction between past and future is a subjective one, dependent upon the existence of conscious observers. Second, it could imply that the distinction is not observer-dependent but perspective-dependent: that each point in time determines a division between past and future. Though not subjective like the first, the second answer is not "objective" either, since the difference between past and future embodied by it is arbitrary, pertaining to the choice of a particular instant on the time axis rather than to the universe that the time-axis measures. One might as well say that there is an objective difference between "left" and "right" on a line, since every point divides the line into left and right segments. On either count, then, Reichenbach's distinction between past and future fails the test of objectivity.

What must now be asked is whether the distinction between past and future of Theory *D*, the dynamic theory, does not likewise fail the test of objectivity. Although there is an obvious difference between the branching future and the unbranched past of every universe-picture, it might be said that the division between the two is arbitrary in the sense that *which* branching future and unbranched past we select for consideration depends upon our choice of a particular frame-time. If we choose 1800 A.D. we get one universe-picture; if 2000 another. Analogously to the image of the line which we divide where we will, we might think of the universe as a bush from which we cut an infinite number of brooms, one for each frame-time. Only the universe is objective: how we cut it up into past and future depends on which temporal perspective we select.

As soon as this analogy is produced, however, it is apparent that it misrepresents Theory *D*. Given that the universe-picture for 1976 has the form of a tree, it is *not* arbitrary how we select the tree for the year 2000. In fact we cannot select it at all. If we recall that there is no distinguished future branch, we see that there is no way in which the universe-picture for 2000 can be constructed in 1976. Hence the situation of Theory *D* is entirely different from that of Reichenbach's theory, which allowed any moment to serve as the division between a "determined" past and an "undetermined" future. Unlike Reichenbach's theory, Theory *D* does not allow an arbitrary moment to constitute a "present", i.e., to be the divide between a branched future and an unbranched past on some universe-tree. The year 2000, for example, cannot constitute a "present" since there exists no means of determining (other than waiting and seeing) which universe-tree it serves to divide. Similarly, the year 1800 cannot serve as "present", since not all of the branches of its universe-tree are on a par: one and only one of them is distinguished as *the* future. If it is objected that the universe-tree which we have spoken of as constituting "the universe" is a different universe-tree from one instant to the next, so that the extensions of "past event", "present event", and "possible future event" change with time, then it can only be replied that this is correct, and reflects the dynamic character of Theory *D*. But it is not arbitrary (as it is in the case of Reichenbach's theory) which moment is selected as "present". Hence even if what we call the present is relative to a time, it is not relative to *the choice of* a time. This account of the present is certainly relational rather than absolute, but the choice of which moment constitutes the present is not arbitrary but forced.

The situation in the case of Theory *D* is quite different from that of Theory *A*, in which, given a complete (four-dimensional) universe-description for any frame-time, one can construct a complete description for any other frame-time by using Lorentz transformations. In contrast, given a complete description *U* of the universe at a time *t* in a coordinate frame *f*, Theory *D* allows for very few Lorentz-type transformations:

(*a*) Where *f'* is any coordinate frame derived from *f* by a rotation of the time-axis, no corresponding transformation of *U* into *U'* exists.[8]

(*b*) Where *f'* is any coordinate frame derived from *f* by translation along the time-axis, a transformation of *U* into *U'* exists when the translation is in the direction of decreasing *t*, but not when in the direction of increasing *t*.

The reason why there exists a method of transforming a later universe-picture into an earlier one within the same frame of reference, but not an earlier one into a later one, is that earlier branches can be "recovered", although lack of a distinguished future branch prohibits construction of future universe-pictures. The lack of Lorentz-type transformations by which to proceed from one universe-picture to another implies that what is meant by "the universe" in Theory *D* is something less independent of temporal perspective—less of an "object" perhaps—than the universe of Minkowski's spacetime. Felix Klein, in his Erlanger Program, defined the non-perspectival properties of a geometrical object as those which remain constant under all rotations and translations of coordinate axes. But what if some of these transformations cannot be performed: or rather, can be performed only with the passage of time? We must

---

[8] Because there will always be branches of *U* which are missing in *U'*, and there exists no means of determining which branches to omit.

conclude from this that the object being described is time-dependent.[9] The universe, then, is inherently characterized by time flow, and this flow is independent of the existence of conscious observers.

4. Further Discussion and Elaboration of Theory *D*

The characterization in the previous two sections of Theory *D*, which purports to model objective time flow, leaves some loose ends dangling and some questions still unanswered. In this section as many as possible of these loose ends will be gathered up.

(*i*) *Does a dynamic universe-model presuppose a second time-dimension?* If the universe has the structure of a tree, and if the difference between earlier and later universe-trees is that the latter have fewer branches, it would seem that the universe is like a tree that "grows" by losing branches. But anything that grows or changes can do so only in time, and since the universe is already a collection of four-dimensional manifolds, a fifth dimension would be needed for it to change in. What this shows is that the analogy with a three-dimensional tree, which grows and changes in time, should not be carried too far. A second time dimension would be an extravagance in a theory that is already extravagant enough. The extravagance can be avoided in one of two ways. We can, if we wish, deny that every entity which is different at different times is an entity which changes.[10] Alternatively, we can say that the progressive falling away of future branches on the universe-tree does not "take" time, but instead "generates" time.[11]

(*ii*) *Branching toward the past.* All the theories considered in Section 2 branch, if they branch at all, toward the future. None branches toward the past. Does the omission of past-branching models reflect some prejudice or bias? The future branches represent all those sequences of states of affairs which are physically possible relative to a given instantaneous world-state. But no preferred time direction attaches to the notion of relative physical possibility: *A* can be

---

[9] Adapting a principle about change found in Earman ([5], p. 264), we might say that if putative time flow is to count as real time flow, then it must not be possible to transform it away. To find out what the universe is going to be like, the only "transformation" possible is waiting and seeing.

[10] The inference:
X is a three-dimensional object
X is different at different times
Therefore *X* changes
can be correct, without remaining correct if we replace "three-dimensional" by "four-dimensional." I owe this observation to Jonathan Bennett.

[11] This alternative was suggested by one of the referees.

physically possible relative to *B,* and yet be earlier than *B.* Hence if we were to represent all the physically possible sequences of states of affairs relative to a given world-state, we would need branching toward the past as well as toward the future.

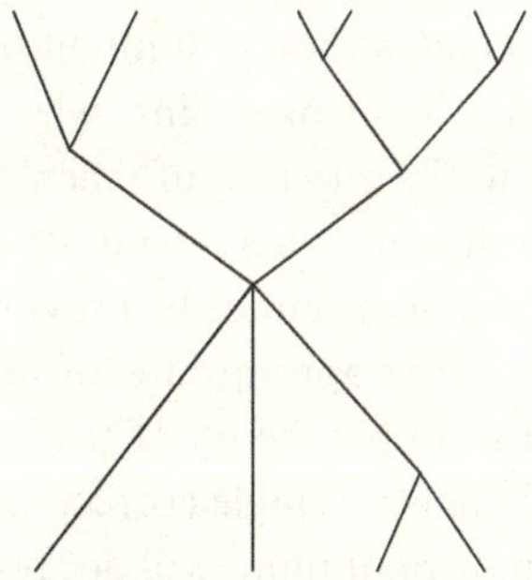

E. Alternative Pasts
Theory

Theory *E*, the theory of alternative pasts, might seem at first sight to exemplify a more consistent approach to universe-building than any of the Theories *B–D.* If possible futures are admitted as a part of complete state-descriptions, what reason, other than an arbitrary one, can be given for the exclusion of possible pasts? Answer: a metaphysical reason reflecting the common belief that the past is unique. Consider any famous unsolved historical problem, say the problem of who wrote the works attributed to Shakespeare. Suppose that conclusive evidence which would settle the problem has by now ceased to exist, so that it cannot be inferred from the present state of the universe whether Marlowe, Francis Bacon or Shakespeare himself wrote them. Will historians conclude that there is no longer any single correct answer to the question of who wrote *Hamlet?* That although there may have been a correct answer in 1601, and at any time up to when the last piece of conclusive evidence ceased to exist, there is none now? No. They will acknowledge that we may never *know* what the correct answer is, but they will cling with every particle of their being to the belief that whatever answer was correct in 1601 will be correct now, and that there cannot have been more than one correct answer in 1601. All of which means to say that historians, like other people, believe in a unique past. We conclude that at most epistemological significance, but no ontological significance, can be attached to a multiplicity of possible pasts.[12]

[12] The exclusion of past-branching universe models implies the impossibility of "changing" the past even in the minimal sense of switching attention from one physically possible past to another. We find traces of the idea that, given a multiplicity of possible pasts, any one of them may be chosen as *the* past, in Borges' story *The Other Death* in [1]. For a speculative view of the

(*iii*) *Arguments for a distinguished future.* It might be thought that the reasoning used above to argue for a unique past could also be used to argue for a unique or preferred future. This argument would hold whether the universe were deterministic or indeterministic. It was stated above that there must be a single correct answer to the question of who wrote *Hamlet*, although it might be impossible for us to know the answer. By parity of reasoning, why should there not be a single correct answer to the question of when a given atom of plutonium will decay, even though it is now impossible for us to know it? A full answer to this problem cannot be provided until the next section, where truth-conditions appropriate for future-tense statements are discussed. In that section a theory of truth will be presented according to which there is no single correct answer to the question of when a given atom of plutonium will decay. For the moment let us simply note once more that Theory *D*, the theory of objective time flow, differs radically from Theory *B* in explicitly denying that any one of the multitude of possible futures is singled out as *the* future, and that this denial goes hand-in-hand with denying that there is an answer to the question of when a given plutonium atom will decay. On some branches it decays at one time, on others at another, and this is all that can be said.

(*iv*) *The relationship between determinism and time flow.* If the universe were deterministic, would a theory of objective time flow be possible? The answer which is consistent with the theory of time flow presented here is in the negative. The reason is, that it is of the essence of Theory *D* that universe-pictures (complete descriptions of the universe) be different at different times, and furthermore that these time-dependent pictures not be "snapshots" from different temporal points of view of some time-invariant reality, but that the universe actually *be* what they describe. This would not be so, however, if the theory of universal Laplacian determinism were true. If universal Laplacian determinism were true it would be possible, given the laws of nature and a complete description of any instantaneous world-state, to construct a complete description of any other world-state, and hence of all world-states. There would therefore be only one universe-picture, valid for all times. The universe would be time-invariant, and any theory of temporal flow would be either subjective, dependent upon conscious experience, or perspectival, resting upon the successive temporal aspects presented by a

---

past as having no reality at all, apart from that portion of it the causal effects of which are still in evidence, see Lukasiewicz [12], p. 38.

time-independent being. Hence determinism is incompatible with objective time flow. This is not to say, of course, that indeterminism implies time flow, since theories *A–C* can all be indeterministic, but only that determinism implies no time flow, when the latter is understood in the "objective" sense.

(*v*) *Possibility and probability.* On the universe-pictures of Theory *D*, each physically possible future sequence of events is represented by a branch. No indication is provided, however, of the relative probability of any one of these branches becoming the "actual" branch. It is far from clear whether any viable account of the probability of future events can be derived from the theory of branching futures, but the model appears to lend itself to the following as a first approximation. Suppose that the half-life of plutonium 239 is 24,400 years, i.e., that there is a 50% chance that a given atom of the substance will decay in that time. This could be represented on a universe-picture by supposing all the branches relative to a given world-state to be equiprobable, and by having the atom decay on 1/2 of them within 24,400 years, on 1/4 of them within the next 24,400 years, etc.

(*vi*) *Fatalism and divine omniscience.* It should be emphasized that nothing said up to now is incompatible with human free choice, or implies that what will happen in the future is in any sense already decided or predestined. To underline this point, two arguments designed to show the predetermining of the future in other than causal ways will be examined. These arguments have the appearance of being based on the kinds of relativistic considerations discussed in Section 3 above, but will be shown to fail to establish their conclusions.

The first of these arguments is a relativistic version of fatalism, put forward by Putnam in [17] and Rietdijk in [21]. Suppose that *O* were an observer on the Crab Nebula, moving in such a way relative to us that the "now" line of his frame of reference intersects the world-line of the earth many years in the future. For *O*, according to what was said in Section 3, the events of the earth's history below his "now" line constitute part of what he regards as the "past"; they form part of the "trunk" of his universe-picture. But if this is so, it might seem that these events were in some sense already "decided" or "predetermined" in a non-causal way, since if they form part of the trunk of his universe-picture they could scarcely avoid, in a few years, forming part of ours. These considerations might appear to support a modern version of fatalism.

It is not difficult, however, to expose the weakness of this argument. Let our frame of reference be f_1, and let *O*'s be f_2. What the argument shows is that relative to a certain time, say time t, in frame f_2, it is "determined" or "true" that, for example, the earth's oil resources are exhausted in the year 2000. That is to say, the universe-picture for the frame-time (f_2, t), if it could be drawn, would show the oil wells dry in 2000 A.D. But of course it *can't* be drawn: there is at the frame-time (f_1, 1976) no universe-picture for the frame-time (f_2, t). We can, to be sure, change our frame, by accelerating ourselves briefly so that our motion parallels that of the Crab Nebula, and this will give us a universe-picture for the frame-time (f_2, t^*), but t^* is very much earlier than t. No transformation (other than waiting and seeing) takes us from the universe at (f_2, t^*) to the universe at (f_2, t), and no meaning can therefore be attached to the statement that the events of 2000 A.D. are "already" prefigured in other observers' universe-pictures.

The second argument concerns God's omniscience. Although we on earth, it might be argued, cannot cognize the universe-picture for the frame-time (f_2, t), God can, and His foreknowledge makes human free choice illusory. The attempt to argue from God's foreknowledge to the illusoriness of choice takes two forms, depending on what doctrine is adopted of the nature of God, but in neither form does it withstand close examination.

If God is considered to be a being outside of and apart from time, then it might well be true that God can cognize all universe-pictures, including that for the frame-time (f_2, t) in which, let us suppose, the oil wells are dry in 2000 A.D. But if God is outside time, then whatever He knows is known not at any particular time, but in a non-temporal way. And in that case there would seem to be no conflict between omniscience and free will, since God does not know what I am about to do *before* I do it, nor what policies are going to be adopted about oil supplies *before* they are adopted.

If on the other hand God is regarded, in the style of process theology, as a being whose essence is at least partly temporal, and who cognizes the events of the world in the successive way that human beings cognize them, then again there is no conflict between omniscience and freedom. At any given moment, God knows everything that is to be known, i.e., every true proposition. But if the outcome of a future decision, say the decision whether or not to use up the world's oil reserves, is represented by a proposition which is today neither true nor false, then God cannot know this outcome. Furthermore, this constitutes no limitation on God's omniscience, for to know is to know what is true.[13] Against this it might be objected that if God is a temporal being, He would have to experience the coming-into-being of events *either* according to the frame-time (f_2, t) of the observer on the Crab Nebula (in which case he might well cognize the future history of the earth), *or* according to our frame-time (f_1, 1976), *or* according to some other frame-time. But, for God, there is no privileged frame-time.[14] Consequently there is no privileged universe-picture. If God

[13] The position that God's omniscience is in no way diluted by His inability to know the truth or falsehood of future contingent propositions was adopted by the 14th-century philosopher Gersonides. See [2], p. 80.

[14] Fitzgerald, in [7], p. 325, makes essentially this point.

cognizes any universe-picture He would seem to have to cognize them all, even the future ones. But this objection fails to take seriously the consequences of asserting that God is a temporal being. If God really does cognize universe-pictures in a successive way then He cognizes each of them at a certain time. Which time? The most natural answer is, at the frame-time of the universe-picture in question. Thus God cognizes the universe-picture of the observer on the Crab Nebula at the frame-time (f_2, t), our universe-picture at the frame-time (f_1, 1976), etc. Divine omniscience, according to this view, consists in knowing all universe-pictures, but in a successive way that avoids any conflict with free will.

5. Two Different Theories of Truth

What has been said up to now about time and becoming has been mainly metaphysical in tone: we have been concerned with presenting a picture of time flow in the physical world. In this section the emphasis will be on logic rather than metaphysics. We shall consider not events but statements about events; not metaphysical theories but truth-conditions for temporal propositions. Our aim is a theory of truth which fits the theory of objective time flow.

We begin by calling attention to what seems at first sight an uncontroversial feature of how we understand and employ the concept of truth. Suppose at time t_0 I utter a true sentence in the present tense:

(1) There are no free parking spaces on Woodmont Street.

If (1) is true at time t_0, at every moment later than t_0 it would appear to be true to say that there *were* no free parking spaces on Woodmont Street, and, at every moment earlier than t_0, that there *will be* no free parking spaces on Woodmont Street. The same holds if (1) is dated:

(2) At time t_0, there are no free parking spaces on Woodmont Street.

If (2) is true at t_0, then either it or one of its tensed variants (past or future tense) is true at any other time. We can if we wish remove the tense from (2) entirely, and, following Smart ([22], p. 133), italicize the verb to indicate that it is to be considered as tenseless:

(3) At time t_0, there *are* no free parking spaces on Woodmont Street.

With (3), we have arrived at a form of words whose truth-value is apparently independent of the time it is uttered: a proposition which, if true at any time, is true at all times. These considerations may induce us to make the following generalization:

(*A*) If a sentence *S* is true at any time, then either *S* or a past or future-tensed variant of *S* is true at any other time.

Generalization (*A*) is the central tenet of what we shall call the theory of timeless truth (cf. [13], p. 271). This theory, which differs from other theories for which generalization (*A*) does not hold, is the natural companion of the mind-dependence theory of time flow. For the objective theory, a more appropriate theory of truth must be devised.

The reason why the theory of timeless truth fits the mind-dependence theory of time flow is that both theories conceive of the real world as a time-invariant entity. For the mind-dependence theory, "the objective world simply *is*, it does not *happen*." For the theory of timeless truth, the element of timelessness about the truth of a true sentence is provided by relating the sentence to something enduring, something which perpetually underwrites the truth either of the sentence itself or of one of its past or future-tense variants. If in examples (2) and (3) above we call t_0 the "time of reference" of the sentence in question, and distinguish it from the "time of assertion", then what the theory of timeless truth says is that, no matter what its time of reference may be, a sentence true for one assertion time is true for all assertion times (allowing for tensed variants). The time of reference ties the sentence to the real world, and the invariance of truth under changes of assertion time provides for its omnitemporality.

When we come to consider the theory of objective time flow, on the other hand, the theory of timeless truth is no longer appropriate. The universe ceases to be time-invariant, and therefore no longer provides a fixed point for true sentences to be related to. Instead of a single universe-picture we have a succession of universe-pictures, each one capable of underwriting the truth of some true sentences but none capable of underwriting them all. What theory of truth replaces the timeless theory? Consider the universe-picture W_{ft} for the frame-time (f, t), and let S be the sentence:

(4) Event *E occurs* at time t_0.

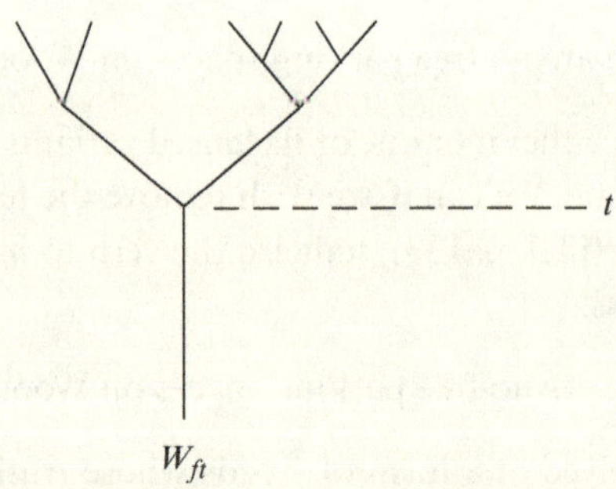

(We shall hereafter refer to sentences of type (4), or to tensed versions of (4), as "event sentences.") In formulating truth-conditions for S at frame-time (f, t), three possibilities present themselves.

(*i*) $t_0 = t$. In this case we seek to know under what conditions *S* is true at a time simultaneous with *S*'s time of reference. Its "assertion time" and its time of reference coincide. The truth-conditions are: If event *E* is to be found at the lowest node of the universe-picture W_{ft}, then the sentence *S* is true. If not, it is false.

(*ii*) $t_0 < t$. Here the assertion time is later than the time of reference. To determine whether *S* is true or false we must look farther down the trunk of the universe-picture. If *E* is located at the level of time t_0, *S* is true; if not, false.

(*iii*) $t_0 > t$. This is the most difficult case. We must formulate the truth-conditions for a sentence whose assertion time is earlier than its time of reference, bearing in mind that no ontological mark distinguishes whichever branch will become actual from its fellows on the universe-picture W_{ft}. In [13], p. 276 it is said that *S* is true if and only if there exists at *t* some condition (set of events) sufficient to make *S* true at the later time t_0. Equivalently, we may say that *S* is true if and only if the event *E* occurs at time t_0 on all branches of W_{ft} (recall that this set of branches represents the set of all physically possible futures). If *E* fails to occur at t_0 on any of the branches of W_{ft}, then *S* is false, and if *E* occurs on some and fails to occur on others then *S* is neither true or false, but has a truth-value that is undefined.

The above set of truth-conditions, which specify what it is for an event-sentence to be true or false at different times, characterizes a theory of truth which differs from the timeless theory. We shall call it the "temporal" theory of truth. In contrast to generalization (*A*) above, the temporal theory asserts:

(*B*) If an event-sentence *S* is true at a certain time *t*, then either *S* or a past-tense variant of *S* is true at all later times. *S* or a future-tense variant of *S* may or may not be true at times earlier than *t*.

According to the theory of timeless truth the law of bivalence, which states that every sentence is either true or false, applies to all event sentences. But according to the temporal theory, the law of bivalence fails for certain event sentences when their assertion times are earlier than their times of reference. In the next section formal semantics will be presented for a language containing event sentences whose truth-conditions follow this pattern.

6. Semantics for the Temporal Theory of Truth

Our language will be a simple one, containing only atomic sentences and truth-functions. The atomic sentences are of the form *Otxa*, to be read "an event of type *a* occurs at time *t* and at place *x*." The notion of an event-type derives from von Wright ([25], p. 36), and is exemplified by such things as "exploding", "colliding", and "buttering a piece of toast". It is of the essence of event-types that they are repeatable; the same event-type can be instantiated at different times and places. In this way they differ from individual events, each of which is tied to a particular time and place. An individual event can be regarded as the instantiation of an event-type, or of a set of event-types, at a particular place-time.

The semantics for our event-language, unlike the language itself, are quite complicated. They would not be worth presenting here in any detail, were it not for the fact that in addition to providing truth-conditions for the event-language, they also serve as a more precise characterization of what a universe-picture is, and of how the universe can be different at different times.

The basic semantic concept required is that of a set U of spacetime points. U is a topological space, the exact properties of which are not yet clear (U is connected and metrizable, although because of its branched character it cannot be a manifold). E is a set of event-types, and g a function which assigns one or more members of E to each member of U.[15] F is a set of coordinate frames, and each $f \in F$ partitions U into simultaneity classes of points in a manner to be described below. T is a set of time coordinates (i.e., real numbers), and h is a function which, for each frame f, assigns each $u \in U$ a time. Similarly X is a set of place-coordinates (triples of real numbers). L is a binary relation on world-states (to be defined below) which determines the precise mode of branching of each universe-picture. k is a "location function" which, given a frame, a time, a place and a branch on a universe-picture, singles out the appropriate member of U. Finally u^* is a distinguished member of U, namely the point "here-now". To sum up, our model structures are ordered ten-tuples $\langle U, E, F, T, X, L, g, h, k, u^* \rangle$, where U, E, F, T, X, L, u^* are as stated above, and

$$g: U \rightarrow \mathcal{P}\mathrm{E}$$

$$h: U \times F \rightarrow T$$

$$k: F \times T \times X \times B_{ft} \rightarrow U,$$

where B_{ft} (to be defined) is the set of branches of the universe-picture for frame-time (f, t).

A frame f partitions U into simultaneity classes as follows. Two points u_1 and u_2 belong to the same simultaneity class S_{ft} if and only if $h(u_1, f) = h(u_2, f) = t$. Because of the branched character of U, the points which belong to a given simultaneity class do not in general constitute a single instantaneous world-state. Instead they divide into many disconnected world-states, each one

---

[15] Or, if some regions of spacetime are "empty," there might be some members of U to which g assigns no members of E.

inaccessible to its neighbors. Thus two members u_1 and u_2 of the simultaneity class S_{ft} belong to the same world-state if and only if u_1 and u_2 are arc-wise connected in S_{ft}.[16] Each world-state is topologically connected, while S_{ft} in general is not. Let W_f be the set of all world-states belonging to all the simultaneity classes into which *f* partitions *U*. W_f, ordered by the relation *L*, has a branched structure, as may be seen from the following diagram. The parallel dotted lines

(4) Event *E occurs* at time t_0.

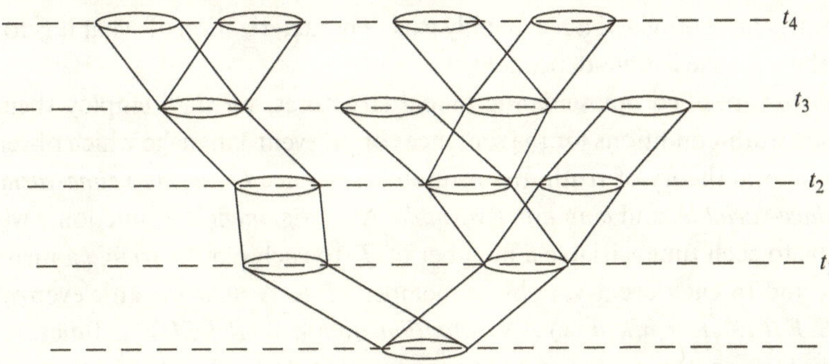

represent the partitioning of *U* into simultaneity classes with respect to the frame *f*, and each ellipse represents a world-state. (On the diagram, *U* is depicted as having an unique world-state at t_0, the first instant, but this cosmological feature of the model can be avoided if necessary. W_f can have a branched structure without having a unique root.)

The element of the semantics which bestows upon W_f its branched structure is the relation *L*. Where w_f and w'_f are elements of W_f, it is appropriate to read $Lw_fw'_f$ as "world-state w_f is a temporal successor of w'_f". On the diagram above, for example, the leftmost world-state of the simultaneity class at t_3 is a temporal successor of the leftmost state of t_2, but not of the middle state at t_2. A *branch* of W_f is defined as a maximal *L*-chain on W_f. W_f's tree-like characteristics are determined by the properties of the relation *L*: in general *L* will be transitive, dense, continuous, and non-branching toward the past.[17]

Although W_f has a tree-like structure, it does not yet correspond to what we have called a "universe-picture". Universe-pictures are different at different times and are shaped like trees with trunks, whereas W_f is time-invariant and has no trunk. The trunk is provided by u^*. Let us suppose that the time is now t^*, where $t^* = h(u^*, f)$. Of all the world-states belonging to the simultaneity class for t^* relative to the given frame *f*, there is one and only one which is the actual world-state of the universe, namely the one containing u^*. Call the world-state

---

[16] That is, connected by a continuous path of points, each of which belongs to S_{ft}.

[17] For an examination of the various constraints that may be placed upon *L*, see [16], pp. 38–57. Non-branching toward the past, for example, requires that $(Lxy \;\&\; Lxz) \supset (Lyz \vee Lzy \vee y = z)$.

in question w_{ft^*}. Now construct the universe-at-the-frame-time (f, t^*) by dropping out all the branches of W_f that do not pass through B_{ft^*}. Call the resultant tree W_{ft^*}, and the set of its branches B_{ft^*}. This, then, will be the entity which in Section 3 we called the universe-at-a-frame-time.

We have as yet no name for the tree W_{ft^*}. Let us call it a *universe-tree.* Universe-trees are what universe-pictures portray. Being dependent upon the choice of u^*, universe-trees are different at different times. If what we call "the universe" is precisely the latest member of a sequence of universe-trees, our semantics have provided a reasonably exact characterization of what it is to say that the universe is time-dependent.

Having specified our semantic model structures, we shall employ them to provide truth-conditions for the sentences of an event-language which observes the temporal theory of truth. In any atomic sentence *Otxa, t* is a *time-variable, x* a *place-variable*, and *a* an *event-variable.* An *assignment* is a function *v* which assigns to each time-variable a member of *T*, to each place-variable a member of *X*, and to each event-variable a member of *E.* A *model* is an eleven-tuple $\langle U,E,F,T,X,L,g,h,k,u^*,v\rangle$. A *valuation over a model M* is a function v_M which gives each sentence of the event-language a truth-value in the model, that is to say one of the three values *true, false*, and *undefined.* I shall sketch briefly a way of introducing trivalent valuations which allows the resulting logic to be classical.

The basic idea of our semantics is that event sentences *Otxa* whose assertion times are earlier than their times of reference should be true if and only if the events they refer to occur in the appropriate positions on all branches of the corresponding universe-tree; that they should be false if the events occur in the right positions on none of the branches; and that they should be neither if the events occur on some but not all branches. Leaving aside cases in which we try to compute the truth-value of sentences uttered by observers at frame-times different from ours (which requires reconstructing their universe-trees), we consider only simple cases in which the assertion time of all event sentences is our own. This assertion time is always t^*, and the relevant semantic structure is the tree W_{ft^*}. Truth-values are relative both to t^* and to the chosen frame *f*:

$v_M(Otxa, t^*, f) = T$ iff for every branch $b_{ft^*} \in B_{ft^*}$ there is a $u \in U$ such that $k(f, v(t), v(x), b_{ft^*}) = u$ and $v(a) \in g(u)$.

In other words, *Otxa* is true in model *M* if on every branch of the tree W_{ft^*} there is a point *u* located at time *t* and place *x*, and if the set of event-types instantiated at *u* contains *a.* Similarly *Otxa* is false if no branch of W_{ft^*} contains such a point, and neither true nor false if some do and some do not. Truth-functional compounds of atomic sentences may be assigned values using van Fraassen's method of supervaluations, for an explanation of which see Thomason's [23].

It might be objected that the truth-conditions laid down for sentences whose assertion times are earlier than their times of reference are extremely strong. This is so. Consider the following example:

(1) The earth *collides* with a comet in the year 2000.

If, as is quite possible, the position and velocity of the earth and the comet in 1976 are such that they are on a collision course for the year 2000, and if no outside influences such as rocket motors mounted on the earth can alter this, then the collision in the year 2000 will be on all possible futures and (1) will now be true. These conditions might seem to have the effect of equating truth about the future with inevitability. This may well be the case: the conditions *are* strict, but probably no more so than everyday life ones for the truth of future-tense statements. The man who appears in the turbinehouse and asks breathlessly if it is true that the dam is going to burst is not inquiring whether the bursting is a feature of only *some* possible futures.

7. Physical Possibility and Physical Law

One last topic, which will be discussed briefly, concerns the question of whether natural laws can he based upon the prior notion of physical possibility. The usual approach would be the other way around. That is to say, the class of physically possible futures relative to a given world-state *w* would be defined as those which, given the laws of nature, were compatible with the initial conditions obtaining at *w*. Here the notion of physical possibility if defined in terms of the notion of physical law: what is physically possible is what is permitted by the laws of nature. There is, however, another way of proceeding, and that is to stand the previous definition on its head and define the notion of physical law in terms of the notion of physical possibility. In this case the primitive idea is that of a class of alternative futures or branches on the universe tree, all of which are physically possible relative to a given set of initial conditions, and which collectively exhaust the range of such possibilities. If a certain event is contained in every such alternative, we define it as being physically necessary relative to the initial conditions. What we call "laws of nature" are generalizations of individual instances of physical necessity defined in this way.

Regarding laws of nature as resting upon physical possibility is not orthodox, but has certain advantages. First, it provides a firm ontological basis for the necessity of natural laws. This ontological basis is the universe itself, which is a branched structure of alternative courses of events. The mode of branching, and the kinds of events contained on each branch, determine what is law-like and what is accidental. Relative to any given set of initial conditions, there is a class of physically possible outcomes or futures, and this class exhausts the range of such possibilities. Long experimentation, perhaps spread out over hundreds of years, in which the initial conditions are reproduced again and again, will gradually reveal what the members of the class are. Since the class in question has a completely "objective" existence, justification is provided for the assertion that the laws of nature are discovered, not made. But secondly, in a certain sense the laws of nature are made, not discovered, for the linguistic and mathematical form in which they are expressed is not dictated by the ontology of possible futures but is a function of the creative element in man. Finally, the interpretation of physical law in terms of physical possibility makes plausible the idea that the laws of nature might change with time. What is physically necessary is what is common to the class of all outcomes relative to a given set of conditions. Although the class in question may be invariant throughout our epoch, in another epoch it might differ, which is to say that what is physically necessary *now* may not be physically necessary *then.* In such a case not only the universe itself, but also its laws, would be time-dependent.

References

[1] Borges, J. L. *The Aleph and Other Stories, 1933–1969*. New York: E. P. Dutton, 1970.

[2] Cahn, S. *Fate, Logic and Time.* New Haven: Yale University Press, 1967.

[3] DeWitt, B. S. "Quantum Mechanics and Reality." *Physics Today* 23 (1970): 30–35. Reprinted in [4].

[4] DeWitt, B. S. and Graham, N. (eds.). *The Many-Worlds Interpretation of Quantum Mechanics.* Princeton: Princeton University Press, 1973.

[5] Earman, J. "Space-Time, or How to Solve Philosophical Problems and Dissolve Philosophical Riddles Without Really Trying." *The Journal of Philosophy* 67 (1970): 259–277.

[6] Eddington, A. S. *Space, Time and Gravitation.* Cambridge: Cambridge University Press, 1920.

[7] Fitzgerald, P. "The Truth about Tomorrow's Sea-Fight." *The Journal of Philosophy* 66 (1969): 307–329.

[8] Grünbaum, A. *Modern Science and Zeno's Paradoxes.* Middletown, MA: Wesleyan University Press, 1967.

[9] Grünbaum, A. "The Anistropy of Time." In *The Nature of Time.* Edited by Gold. Ithaca: Cornell University Press, 1967. Pages 149–186.

[10] Grünbaum, A. "The Meaning of Time." In *Basic Issues in the Philosophy of Time*. Edited by Freeman and Sellars. Lasalle: Open Court, 1971. Pages 195–228.

[11] Grünbaum, A. *Philosophical Problems of Space and Time.* Second edition. Dordrecht: D. Reidel Publishing Co., 1974.

[12] Lukasiewicz, J. "On Determinism." In *Polish Logic 1920–1939*. Edited by S. McCall. Oxford: Clarendon Press, 1967. Pages 19–39.

[13] McCall, S. "Temporal Flux." *American Philosophical Quarterly* 3 (1966): 270–281.

[14] McCall, S. "Time and the Physical Modalities." *The Monist* 53 (1969): 426–446. Reprinted in *Basic Issues in the Philosophy of Time*. Edited by Freeman and Sellars. LaSalle: Open Court, 1971. Pages 102–122.

[15] McCall, S. "The Cardinality of Possible Futures," (abstract), *The Journal of Symbolic Logic* 35 (1970): 363.

[16] Prior, A. N. *Past, Present and Future.* Oxford: Clarendon Press, 1967.

[17] Putnam, H. "Time and Physical Geometry." *The Journal of Philosophy* 64 (1967): 240–247.

[18] Reichenbach, H. "Les Fondements Logiques de la Mécanique des Quanta." *Annales de l'Institut Henri Poincaré* 13 (1953).

[19] Reichenbach, H. *The Direction of Time.* Berkeley: University of California Press, 1956.

[20] Rescher, N. *The Primacy of Practice.* Oxford: Blackwell, 1973.

[21] Rietdijk, C. W. "A Rigorous Proof of Determinism Derived from the Special Theory of Relativity." *Philosophy of Science* 33 (1966): 341–344.

[22] Smart, J. J. C. *Philosophy and Scientific Realism.* London: Routledge and Kegan Paul, 1963.

[23] Thomason, R. H. "Indeterminist Time and Truth-Value Gaps," *Theoria* 36 (1970): 264–281.

[24] von Wright, G. H. *Norm and Action.* London: Routledge and Kegan Paul, 1963.

[25] Weyl, H. *Philosophy of Mathematics and Natural Science.* Princeton: Princeton University Press, 1949.

{ 17 }

Time Flow

Majority opinion among scientists and philosophers of science is against time flow or temporal passage as a property of the physical world. Eddington, writing in 1920, gives an eloquent and poetic statement of the scientific position:

> Events do not happen; they are just there, and we come across them. "The formality of taking place" is merely an indication that the observer has on his voyage of exploration passed into the absolute future of the event in question (Eddington (1920), p. 51).

The alternative to considering temporal becoming as a physical process is to make it a subjective phenomenon, as Adolf Grünbaum does in putting forward the "mind-dependence" theory.[1] Grünbaum is careful to distinguish between temporal *anisotropy* and becoming. To say that the four-dimensional world is anisotropic along the time axis is to say that there exist processes which take place either normally or invariably in one of the two temporal directions, and rarely if ever in the other. In four-dimensional language, there exist four-dimensional volumes corresponding to these processes which exhibit a characteristic structure or "grain" in the temporal dimension, this grain having a uni-directional pattern. For example, hot and cold water mix to form lukewarm water; lukewarm water almost never "unmixes" into hot and cold components. A match lit on a dark night emits concentric expanding spheres of photons; rarely if ever do concentric shrinking spheres converge on a single point.

But the anisotropy that the world exhibits along the time axis, of which these are examples, does not by itself serve to pick out one of the two temporal directions as THE direction of time. It is arguable that only the property of flow or becoming can do this. If time flows, the direction of time is of necessity the direction in which it flows.

---

[1] Grünbaum (1963, pp. 324–325).

This latter point requires elaboration. If there is such a thing as temporal passage or flow, then it is a matter of logic that the flow is *from* the past *to* the future. It would be absurd to suppose otherwise, and postulate that time might flow from future to past. Hence if there is time flow its existence *ipso facto* (i) divides the time axis into a past end and a future end, and (ii) distinguishes the past-to-future direction as THE direction of time, in opposition to the future-to-past direction.

Neither of these things can mere anisotropy accomplish. If tomorrow we entered into a period in which lukewarm water habitually separated into hot and cold components, or in which entropy habitually decreased, this would not by itself imply that time had changed direction. Hence time flow, if there is such a thing, is logically stronger than temporal anisotropy. None of this goes to show that flow or becoming exists, but only indicates the area in which the truly distinctive character of time resides.

Let us return for a moment to Grünbaum's mind-dependence theory. For Grünbaum, a universe which contained no conscious beings would exhibit no temporal becoming. In such a universe, things would have temporal extension, i.e. would be of a certain age as measured by the difference between their time coordinate in some reference frame and the time coordinate of their origin. But nothing in the strictest sense would "grow older", or "become larger", or change in any way with the passage of time, since in a universe without conscious beings there would be no passage of time. Something X's "becoming larger" in a four-dimensional world means only that later stages of X's world line have broader cross-sections than earlier stages. For Grünbaum, temporal passage or flow is a type of subjective illusion, in the same category as Locke's secondary qualities of colour, taste, and smell.

This is a perfectly consistent philosophical position, one that coheres elegantly with relativity theory. In Minkowski's formulation, special relativity requires replacing the familiar world of three-dimensional objects changing in time by a four-dimensional world in which all "changes" are represented by the interweaving and alteration in cross-section of the world lines of four-dimensional individuals. Thus a given four-dimensional volume may have a three-dimensional cross-section at one time which has the shape of a frog, and another cross-section at an earlier time which has the shape of a tadpole. Change in the four-dimensional world is no more than this.

The four-dimensional world is static not dynamic; it represents change within it, but does not itself change. In the (unbranched) four-dimensional world, time most emphatically does not flow. Hence Grünbaum's theory, which locates time flow wholly within the domain of conscious experience, appears to be ideally suited to the world-view of modern physics.

Although not endorsing time flow, modern physics does not rule it out as a logical impossibility. Its attitude is empirical rather than logical; time flow may exist, but (i) there is no hard experimental evidence for it, and (ii) it plays no role in physical theory. A stronger position is taken by those philosophers who follow McTaggart in claiming that the whole notion of things or events changing their temporal status, from "future" through "present" to "past", is literally absurd.[2] For these philosophers there not only *is* no such thing as time flow; there *could not* be. But, as will be shown, McTaggart's alleged proof of the impossibility of temporal becoming rests upon an error.

The proof runs as follows. The distinctions "past", "present" and "future" apply to events, and they are incompatible. Every event must have one or the other, but no event can have them all. Nevertheless, McTaggart claims, every event *does* have them all. If an event *e* is past (*Pe*), it has been present (*Ne*); if present, it was future (*Fe*). For any event *e* we have *Pe* and *Ne* and *Fe.* But these three are incompatible. To suppose that they are all true is a contradiction.

The obvious reply is that *e* is not *simultaneously* past, present and future, but e.g. *is* present, *was* future, and *will be* past. But according to McTaggart this move to compound tenses will not help, for no event can have all nine of the compound tenses *PP, PN, PF, NP, NN, NF, FP, FN, FF.* In particular, *PP* and *FF* are incompatible. Every attempt to get rid of the incompatibility by moving to more complex compound tenses is met by a further incompatibility at the higher level.

This argument is a sophism. At *no* level do all the tensed attributions simultaneously hold of any event *e*, so that although *PPPe* and *FFFe* are incompatible no one ever would, or could, truly assert more than one of them at a given time. "X is hot" and "X is cold" are incompatible, but no contradiction results if they are not asserted about the same thing at the same time. As will be seen below, there exists an ontological model of time flow which provides a clear proof of the consistency of the concepts "past", "present" and "future", and of the notion of "temporal passage", in the same way that the surface of a sphere provides a clear proof of the consistency of Riemann's elliptic variety of non-Euclidean geometry, or in the way that Tarski's semantics prove the consistency of first-order logic. Hence McTaggart's proof cannot succeed.

McTaggart's views on time, in particular his distinction between the A-and the B-series, have led to an extensive study of temporal language, including (i) the role of tenses and temporal indexicals, and (ii) the question of whether everything that can be said using the latter can also be said using tenseless dated forms. Thus "There will be an explosion" is sometimes analysed as "There *is* (tenselessly) an explosion at time *t*, and *t* is (tenselessly) later than this utterance". The general consensus regarding such de-tensing is that it works with one exception. Tenseless equivalents of almost all tensed utterances can be produced, the exception being tensed statements of the form "It is now six pm, December 29th".[3] This is an interesting and possibly significant result, which

[2] McTaggart (1908), (1927, ch. 23); Dummett (1960); Mellor (1981); Horwich (1987, pp. 18–25).

[3] Smith (1994).

I shall not pursue here. Another major development in the study of temporal language has been the emergence of formal tense logic.[4]

None of these results concerning language, however, touches upon the central issue of this chapter, which is the ontological status of becoming. Temporal language is important because we want to say things about time without contradiction and without absurdity. But the primary issue is not merely to avoid contradiction, important as this may be. It is whether temporal becoming is a feature of the world, i.e. whether there exists such a thing as the passage of time. If there is, we must of course be careful of the language we use to describe it. But to give linguistic issues priority, and try to draw physical and ontological conclusions from them, is to put the cart before the horse.

So much for preliminaries. I mentioned earlier that for physics, the existence or non-existence of time flow is an empirical matter. McTaggart to the contrary, temporal becoming is not a logical absurdity, and if one day the concept of becoming came to play a role in physical theory analogous to that of "field", or "force", or "virtual particle", or "event", then its existence would be recognized as part of the theoretical framework of science. Nothing rules this out. My object in this chapter is to contribute to the process of recognition by showing that a clear and consistent spacetime *model* of time flow is constructible, a model to which the world we live in could conform. The construction of the model will be rigorous and precise, so that any vagueness about becoming (so modelled) which remains in the mind of the reader will be my fault, not the reader's. Besides time flow, many philosophical concepts such as *probability, law of nature, objective chance, quantum non-locality* and *collapse, transworld identity, deliberation* and *decision*, correspond in a natural way to features of the model, the explanatory powers of which are described elsewhere.[5] In what follows I focus on the passage of time, on what it means for events to "happen". Needless to say, happening or becoming will not (in the model) depend in any way on conscious experience, or indeed on the existence of conscious or living beings. Unlike Grünbaum's theory the "becoming" modelled in this chapter is entirely objective, was a feature of the world long before life evolved, and will continue to be a feature of it long after living beings have become extinct.

Briefly described, the model is this. We are to think of the world as a huge four-dimensional collection **W** of point-events. **W** includes the Minkowski world of special relativity but it also includes a great deal more: not only the actual events of the world but all those that are physically possible as well, relative to conditions prevailing at the time of the universe's origin. The events in **W** are arranged in subsets called branches (defined below). **W** as a spacetime is distinguished from all other spacetimes, for example the spacetime of general relativity, by virtue of containing pairs of "inaccessible" events. x and y are

---

[4] Prior (1967); Øhrstrøm and Hasle (1995).

[5] McCall (1994), (1995a), (1995b).

mutually inaccessible if there is at least one branch that contains x but not y, or y but not x.

In **W**, the passage of time is modelled by progressive branch loss. **W** has a complex densely branching structure which it is almost impossible to visualize, but there are orderly subsets W_f of **W** (the parameter f denoting frame of reference) which have a recognizable tree-like shape. Temporal passage is modelled by progressive branch attrition in each sub-model W_f. Time flow, or branch loss in a frame, will be defined once we succeed in isolating the subset W_f from **W**. But if we wish to change frames and represent time flow in a different frame f', then we must go back to **W** and construct the new subset $W_{f'}$. How exactly this is done will be described in what follows.

Besides **W**, the only other spacetime I know of with the topological feature of inaccessibility is Belnap's "Branching space-time" (BST), to which the formal exposition of this chapter owes much.[6] In BST, and in **W**, the fundamental theoretical concept is a binary relation ≤ defined on W, where $x \leq y$ is read "x is weakly prior to y", or "x weakly precedes y". Strict precedence $x < y$, which corresponds to Einstein's "causal relation", is defined as "$x \leq y$ and $x \neq y$". The relation ≤ is a partial ordering on **W**. If we assume that **W** is a topological space, all the other structural concepts of W are definable in terms of ≤. I shall also sometimes read $x \leq y$ as "y is an upper bound of x".

Df. 1 Two points x and y are COMPARABLE if $x \leq y$ or $y \leq x$; otherwise INCOMPARABLE

Using the notion of comparability we can define the concepts of "timelike path" and "lightlike path" in **W**. We first define a more general concept, that of a one-dimensional subset of W known as a "maximal chain".

Df. 2 A MAXIMAL CHAIN in W is a maximal set of pairwise comparable points.

(The "maximality" of a chain C consists of C being both "full" of points of **W** and also being maximally long. A maximal chain is never a proper subset of any other chain.)

Df. 3 A maximal chain is LIGHTLIKE or NULL if for any two points x and y on it where $x < y$, every intermediate point z such that $x < z$ and $z < y$ is already a member of the chain.

A lightlike chain in **W** is a set of points along the path of a light ray, i.e. a set capable of being occupied by a light ray, whether or not a ray is actually occupying them. The set of all null chains or paths through any point x defines x's double light cone, one lobe of the cone being "past" and the other "future."

[6] Belnap (1992). BST, however, lacks the feature of branch attrition.

Df. 4 A TIMELIKE PATH is a maximal chain no part of which is lightlike.

In relativity theory, no object with non-zero mass can travel at the speed of light. The "slower" paths which constitute the world lines of bodies with mass are all timelike, not lightlike.

We now proceed to define "branch" and "inaccessibility."

Df. 5 An UPWARD DIRECTED subset of **W** is a set, any two members of which have a common upper bound in the set.

Thus for any x,y in an upward directed set there is a z in the set such that $x \leq z$ and $y \leq z$.

Df. 6 A BRANCH is a maximal upward directed subset of **W.**

Again, the "maximality" of a branch lies in the impossibility of adding any additional points to it which are not already members of the branch. Every point of **W** belongs to at least one branch. Every pair of branches intersects. If the temporal extension of the world is finite, with a last moment, then some points belong to only one branch. If time extends infinitely, then every point belongs to the intersection of two or more branches. **W** branches only forwards in time, not backwards.

We now define accessibility and inaccessibility.

Df. 7 Two points are MUTUALLY ACCESSIBLE if they share a common upper bound, and MUTUALLY INACCESSIBLE if they have no common upper bound.

Df. 8 A set is ACCESSIBLE if it is pairwise accessible, i.e. if any two of its members are mutually accessible.

Alternatively, a set is accessible if it is a subset of some branch of **W.** We now define a "slice", which is an infinitely thin three-dimensional hypersurface of **W**, only one instant thick, no portion of which is either earlier or later than any other portion. A slice in effect defines a total "world-state" of a branch of **W** at an instant.

Df. 9 A SLICE is a maximal accessible set of pairwise incomparable points.

The next concept to be defined is that of a choice point.

Df. 10 x is a CHOICE POINT if there are at least two distinct timelike paths A and B on which x lies, and if those portions of A and B which are each later than x are pairwise inaccessible.

The existence of choice points indicates *that* **W** branches, but does not determine *how* it branches. In fact **W** branches along "corrugated surfaces", which are, like slices, maximal three-dimensional hypersurfaces only an instant thick, but which have everywhere the slope of a light ray. Imagine a discrete set S of pairwise incomparable choice points, each one separated from its neighbors in S by non-choice-points. Imagine also the set of all future light cones of these choice points, each cone intersecting its neighbors. If we discard all portions of these cones which are later than points at which they intersect with neighboring cones, what we are left with is the "corrugated surface" determined by the set of choice points.[7]

Choice points do not occur randomly in **W**, but are distributed in an orderly way along parallel branching timelike paths. With the exception of "main choice points" (MCP's), which are the limit points of upwardly bounded sets of choice points in a path, sets of choice points are discrete. Thus each non-MCP choice point has a unique predecessor in the path and two or more immediate successors, each one on a different branch. Non-MCP choice points are separated by extremely small temporal intervals. On any given path, main choice points are separated by intervals of constant length Δt of the order of 10^{-24} sec., this distribution being relevant to the definition of objective probability values in W.

We now define the parallel sets of corrugated surfaces along which the well-behaved subsets W_f of **W** branch. Every frame *f* defines a *foliation* of **W**, that is to say a partition of **W** into equal-time hyperplanes with respect to the frame *f*. These hyperplanes are separated into distinct equal-time slices on different branches, it being understood that the partitioning of **W** cuts across branches. Consider an equal-time hyperplane H_f, which cuts across many branches and has sections in each branch. The set of choice points on or immediately above H_f determines a non-connected set of corrugated surfaces C_f along which **W** branches. The complete set of surfaces along which W_f branches is a subset of the set of all corrugated surfaces parallel to C_f. Similarly, in a different frame *f'*, a hyperplane $H_{f'}$, defines the complete set of corrugated surfaces along which $W_{f'}$ branches.

For each frame *f'* a submodel W_f of **W** has now been defined. These submodels have a treelike structure, with branching towards the future and no branching towards the past. *Branch attrition* in each W_f consists in the progressive "pruning" or loss of branches, it being always the lowest branches which disappear. This progressive branch loss is what constitutes, in the branched model, the flow of time.

Suppose for example that the lowest branches in submodel W_f split apart along the corrugated surface C_f at time *t* in frame *f*. Then at time *t* all but one of these lowest branches vanish, leaving the remaining branch to form part of the trunk (the "past") of W_f. If the next surface along which W_f splits is located

[7] A more formal definition of corrugated surfaces is given in McCall (2000).

at time t' of frame f, then all but one of *those* branches are lost at t', and so on. This process of progressive branch loss takes place in every submodel W_f, and makes for objective, non-mind-dependent time flow in each corresponding frame of reference f.

To sum up, there exists in the branched model a way of defining time flow that is both mind-independent and precise. The proposed definition is one which gives becoming or temporal passage a place within the theoretical structure of physics. Conceived of as branch attrition within the overall spacetime structure of the universe, time flow emerges as a physical process unlike any other physical process, but nonetheless one that characterizes the physical world in a real and objective way.

References

Belnap, N. (1992), "Branching Space-Time," *Synthese 92*: 385–434.

Dummett, M. (1960), "A Defence of McTaggart's Proof of the Unreality of Time," *The Philosophical Review*, *69*: 497–504.

Eddington, A.S. (1920), *Space, Time and Gravitation* (Cambridge: Cambridge University Press).

Grünbaum, A. (1963), *Philosophical Problems of Space and Time* (New York: Knopf).

Horwich, P. (1987), *Asymmetries in Time: Problems in the Philosophy of Science* (Cambridge, MA: MIT Press).

McCall, S. (1994), *A Model of the Universe: Space-Time, Probability, and Decision* (Oxford: Oxford University Press).

McCall, S. (1995a), "Time Flow, Non-Locality, and Measurement in Quantum Mechanics," in Savitt (ed.), *Time's Arrows Today* (Cambridge: Cambridge University Press): 155–172.

McCall, S. (1995b), "A Branched Interpretation of Quantum Mechanics Which Differs from Everett's," in Marion and Cohen (eds.), *Quebec Studies in the Philosophy of Science* (Dordrecht: Kluwer): 229–241.

McCall, S. (2000), "Towards a Lorentz-Invariant Theory of Collapse," in Duffy and Wegener (eds.), *Recent Advances in Relativity Theory* (papers presented at a conference at the Imperial College, London): 111–122.

McTaggart, J. M. E. (1908), "The Unreality of Time," *Mind 17*: 457–474.

McTaggart, J. M. E. (1927), *The Nature of Existence*, vol.*II* (Cambridge: Cambridge University Press).

Mellor, D. H. (1981), *Real Time* (Cambridge: Cambridge University Press).

Øhrstrøm, P. and Hasle, P. (1995), *Temporal Logic: From Ancient Ideas to Artificial Intelligence* (Dordrecht: Kluwer).

Prior, A. N. (1967), *Past, Present and Future* (Oxford: Oxford University Press).

Smith, Q. (1994), "The Truth Conditions of Tensed Sentences," in Oaklander and Smith (eds.), *The New Theory of Time* (New Haven: Yale University Press): 69–76.

{ 18 }

QM and STR: The Combining of Quantum Mechanics and Relativity Theory

Bohr would always go in for this remark, "You cannot really explain it in the framework of space and time." By God. I was determined I was going to explain it in the framework of space and time.

—John Slater (quoted in Cushing 1994, 109)

1. Introduction

The aim is to show that quantum mechanics and special relativity can be consistently combined. This requires that the dynamical evolution of a quantum system be describable in any inertial frame, that transformations from one frame to another be admissible, and that quantum mechanical states and laws remain invariant under these transformations.

Many difficulties stand in the way of achieving this reconciliation. To begin with, the state of a QM system is represented by a vector or a density operator in Hilbert space, or by a wave-function in configuration space, and the evolution of the quantum state takes place within that arena. But none of these vector spaces is the familiar spacetime to which the principles of special relativity apply. The first requirement in bringing together QM and STR is to find a way of describing quantum change not in terms of the evolution of the state vector in Hilbert space, but in terms of something changing in ordinary space and time.

Second, collapse is notoriously difficult to picture as a spacetime process. A particle passing through a slit undergoes diffraction, and the wavefunction which specifies the probability of locating the particle in different regions of space spreads out. But when a measurement is made the wavefunction collapses instantaneously, e.g., to a darkened region on a photographic plate. Similarly, a measurement performed on one of two particles in an entangled spin-state partially collapses the superposition and changes the state of the other particle. How are these collapses to be pictured in spacetime? If instantaneous, relative to which reference frame? What mechanism produces an instantaneous distant effect? The most difficult part of constructing a spacetime picture of collapse is dealing with its instantaneous, nonlocal character.

To be sure, one way of treating collapse is to eliminate it and adopt a no-collapse theory; either (i) Bohmian dynamics (Bohm 1952; Albert 1992, Ch. 7; Cushing 1994, 1996), or (ii) the many-worlds interpretation (DeWitt and Graham 1973). These no-collapse interpretations, though of great interest and merit, stand somewhat apart from the present work. The difference is that the latter emphasizes the *probabilistic* character of QM. Neither Bohm's dynamics nor the Everett interpretation allows for objective quantum probabilities; Bohm's because it is deterministic and Everett's because every outcome of every QM experiment is realized on some branch, and consequently there is no sense in speaking of the "probability" of it being realized.

Although in this chapter collapse is held onto, and no-collapse theories set aside, no attempt will be made to give a theory of what causes or "triggers" collapse. This might be an act of consciousness (Wigner), or a GRW-type spontaneous localization process, or something else. All that is provided here is a spacetime model of collapse which is instantaneous and nonlocal, and which is compatible with a variety of different theories of what causes it.

2. The Spacetime Arena

Two questions confront us. First, how is it possible to represent, in space and time, an indeterministic process with more than one possible outcome? Second, what would constitute a spacetime description of the evolution of quantum states, i.e., their temporal evolution in ordinary space as distinct from Hilbert space? Answers to these questions will establish the quantum spacetime arena in which relativistic transformations apply.

Suppose that a single photon passes through a calcite crystal equipped with detectors. To this stochastic process there are two possible outcomes, "+" and "–". The usual spacetime representation of such a process would record only the outcome that actually occurs. The "missing" outcome, the one that was not realized, would not be represented.

An alternative way of representing a two-outcome stochastic process—the one adopted here—is to put in *both* possible outcomes, but to locate them in disjoint spacetime regions which exclude one another. From the common region containing the initial conditions two different regions branch off. This is the spacetime structure at the start of the experiment, when the photon is in a superposition $|+\rangle + |-\rangle$ of plus and minus polarization states. The collapse of the superposition is represented by the vanishing of the branch containing the unrealized outcome. The other branch remains, bringing it about that the experiment has a unique outcome. Spacetime diagrams depicting branching and "branch attrition" of this kind are found in McCall 1994, 1995.

An important advantage of this representation over a nonbranching single manifold representation, in which only the actual outcome is represented, is that branching provides a physical basis for probability values. The values are not attached in some unclear way to each branch, i.e., "put in by hand", but are part and parcel of the overall branching structure.

If for example the photon entering the calcite crystal is already prepared in a pure polarization state oriented at 30 degrees to the crystal's axis, it will have a probability of 75% of passing "+" and 25% of passing "−". These probabilities are built into the spacetime structure in that 75% of the branches contain the "+" outcome and 25% the "−" outcome. If there is a democracy of branches, with each branch having an equal chance of being selected, the corresponding probabilities of "+" and "−" outcomes will be 0.75 and 0.25 respectively. These are objective single-case probability values, uniquely determined by the structure. The probability of a given outcome X is fixed by the proportion of branches on which X occurs, given that each branch has an equal probability of being selected.

For probability values which are rational numbers, finite sets of branches suffice. Thus an outcome with probability *n*/*m* will be located on *n* of a total of *m* branches, relative to a given initial segment. But to represent irrational probability values, such as $\cos^2 20°$, or to represent a stochastic process in which there are infinitely many possible outcomes, as in scattering experiments, radioactive decay times, etc., fixed and definite proportions within infinite sets of branches are needed. (For details see McCall 1994, 88–92.)

Given the spacetime representation of quantum probabilities, the road to defining and representing quantum states is open. What is a quantum state? Unlike a classical state, which assigns a definite value of a physical quantity to a system, a quantum state assigns only an expectation value. More precisely, the quantum state of a system is an exhaustive specification of the probabilities of all possible outcomes of all possible tests that can be performed on the system. It is, in Abner Shimony's words, a "compendium of possibilities", a precise specification of a system's dispositional powers. What must be seen is how this "compendium" can be represented in space and time, and furthermore how it can be represented as changing and evolving.

Imagine as a concrete example a system S to be an electron moving towards an adjustable Stern-Gerlach apparatus which can be tilted so as to measure the electron's spin at any one of 100 different angles $j_1, j_2, \ldots, j_{100}$ To each angle j_i there corresponds an observable O_i, i.e., j_i-spin, and for each O_i the quantum state of S assigns a probability to the values "up" and "down" of j_i-spin on S. In four-dimensional spacetime, the worldline of the electron S branches to couple with each of the 100 possible Stern-Gerlach instruments available to the experimenter by twisting a dial. There is only one three-dimensional instrument, and the four-dimensional representation of the experimental set-up will include one type of branch for each setting.

In the spacetime representation of S's quantum state, each branch associated with the measuring-instrument settings divides again, at a time slightly later than that at which the electron enters the magnet. This branching determines, by proportionality, the probability of the different outcomes spin-up and spin-down for each setting, and it is in these proportions that the quantum state of S lies. If, for example, the electron on entering the magnet is in the pure state $|x+\rangle$, spin-up along the x-axis, then the proportions of "up" and "down" outcomes for each of the 100 observables will take a definite set of values. In branched spacetime. the state $|x+\rangle$ of S consists of these proportions. But if the electron is prepared in the state $|z+\rangle$, or in the mixed state $1/2(P_{x+} + P_{x-})$, the proportion of "up" and "down" outcomes for each of the 100 observables will vary. It is this variation in proportions which constitutes the difference, in four-dimensional spacetime, between the quantum states $|x+\rangle$, $|z+\rangle$, and $1/2(P_{x+} + P_{x-})$.

In the spacetime representation of a probabilistic process, there is one additional element and that is collapse. In the example just described there will be an infinity of spin-up and spin-down branches for each of the 100 observables. In configuration space, the wave-function of the electron in the position basis gets correlated with its spin, and exhibits two peaks in the "up" and "down" channels of the apparatus, the amplitudes of which are functions of the quantum state. But in the end one and only one detector is triggered, the wave-function collapses, and the experiment has a unique outcome. Its uniqueness, in branching spacetime, is brought about by branch attrition, consisting of the survival of exactly one branch out of the infinity which existed at the start of the experiment.

To be accurate, there are two "collapses" in the course of the experiment, the first when the experimenter selects the angle of the measuring instrument, and the second when a single branch gets randomly selected from the infinity of "up" and "down" branches. The first of these is "observer's choice" (Heisenberg) and the second "nature's choice" (Dirac) (see Bohr 1949, 223). In the example considered the role of the observer in selecting the quantity to be measured is an active one, though it is possible to think of situations in which the selection of the measured observable is itself the result of a non-anthropic probabilistic process.

With the identification of a spacetime correlate of quantum states, the first and most important step has been taken in constructing a spacetime arena in which the histories of quantum systems unfold. The quantum state of a system has been identified with the relative numerical proportionality of sets of branches containing all possible outcomes of all possible tests that can be performed on the system. Defined in this way the evolution of quantum states takes place not within Hilbert space but within four-dimensional spacetime.

Continuing with the previous example, suppose an additional option is now available in which the experimenter can route the electron through a magnetic field which changes the orientation of its spin. In Hilbert space, the state vector undergoes a unitary transformation. If now the electron, in its new spin state, is once more confronted with the 100 tests, the up/down probabilities in each case will be altered. This change in the electron's state between t_1 and t_2, i.e., before and after its spin has been rotated, is represented in the spacetime arena as follows.

At t_1 there are 101 branching life-line paths in spacetime for the electron, and its quantum spin state is determined by the up/down probabilities at the ends of the paths leading to the 100 tests. The 101st path, along which the electron's spin precesses, leads once more to the 100 tests at time t_2, and again the electron's life-line branches, but at the end of each branch the proportions of "up" and "down" branches are different from those on the paths that branched off at t_1. In the three-dimensional world, this situation is described by saying that at t_1 the electron was disposed to pass "up" and "down" in each test with one set of probabilities, whereas at t_2 its probabilities of responding were different. Between t_1 and t_2 the electron's quantum state has changed, and the branching structure permits that change to be described in spacetime.

A final point should be made concerning quantum states in Hilbert space and in real space. In Hilbert space there is in principle no limit to the number of different Hermitian operators, for the values of which a given quantum state yields probabilities. But in real space, in any experimental situation, there is at most a finite number of different alternative tests that can be performed on a system. Hence the number of different probability distributions that are built into real spacetime will be only a fraction of those generated by the projections of the state vector in Hilbert space. What this implies is that the quantum state of a system in spacetime has an empirical dimension that the quantum state in Hilbert space lacks. There is no theoretical limit to the number of tests that can be performed on the system, but in practice they cannot all be performed at once.

3. Quantum States Represented in Spacetime and Their Relativistic Transformations

In QM dynamics the state vector $v(t_1)$ at time t_1 is transformed into the vector $v(t_2)$ at time t_2 by the Schrodinger equation $v(t_2) = U(t)v(t_1)$, where $t = t_2 - t_1$. Use of the words "at time t_1" and "at time t_2" implies a reference frame. If the state of a particle p is represented by a wave-function in the position basis which yields probabilities of detecting the particle in different regions of real space, then in spacetime the state of p at time t in frame f will consist of the distribution of these probabilities along an equal-time hyperplane in the frame f (hereafter referred to as a "f-hyperplane"). In a different inertial frame f', states of p consist of probability distributions along f'-hyperplanes, slanted with respect to those in f.

A basic assumption of relativity theory is that if an object or event is describable in one inertial frame of reference then it is describable in all frames, and that there should be rules for translating the description from one frame to another. In branching spacetime, p's state in frame f is translatable into p's state in frame f' by switching to branch proportions along the new family of f'-hyperplanes, these proportions being built into the spacetime structure. If however p's wave-function collapses at or near the time at which such a translation is attempted, there exist well-known difficulties which make the translation impossible. An example from Fleming 1988 makes the problem clear.

In figure 1, a particle p enters into a state reducing interaction (e.g., a position measurement) in a region R of spacetime.

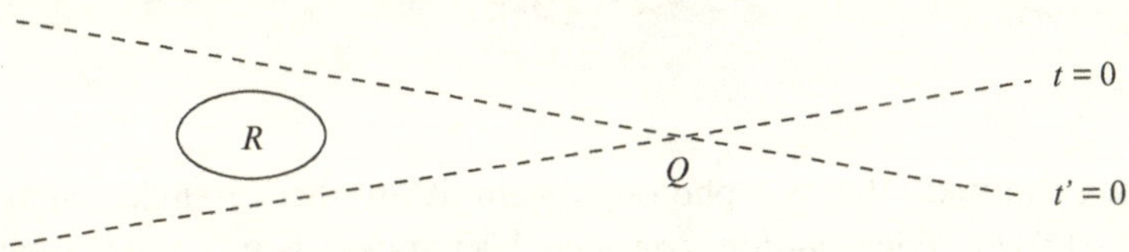

FIGURE 1

Consider the state Ψ of p along $t = 0$ in f, and its state Ψ' along $t' = 0$ in f'. How is Ψ to be transformed into Ψ'?

Relative to $t = 0$, before the position measurement at R, the wave-function of p has a nonzero amplitude at Q. Relative to $t' = 0$, however, p's wave-function either has nonzero amplitude at Q if p is not detected at R, or has zero amplitude if p is detected at R. There is therefore no straightforward way of transforming Ψ into Ψ', since the transformation depends on the stochastic and unpredictable result of the reduction at R.

Fleming's example shows that no transformation "across a collapse" is possible, i.e., a transformation from a hyperplane relative to which a collapse is future, to one relative to which the same collapse is past. However, if transformations across collapses are avoided and if a whole segment of a system's history in one reference frame, including collapses, is translated into a corresponding segment of its history in another frame, also including collapses, then relativistic covariance is preserved. In each frame f, collapse of a quantum state brought about by a measurement will be instantaneous along an f-hyperplane.

I turn now to a second example in Fleming 1988 (discussed extensively in Maudlin 1994, 1996; Fleming 1996; Wayne 1997). Figure 2 (see Fleming 1988,

117; Maudlin 1994, 205) depicts the history of a Bell-EPR experiment using photons, written from the standpoint of two different frames *k* and *k*′.

The sequence of *k*-hyperplanes $t = 1, t = 2, \ldots$ represent moments of the system's history in *k*, and the sequence $t' = 1, t' = 2, \ldots$ moments in *k*′. As Fleming says, we may imagine that *k*′ is the rest-frame of an observer on the ground when the experiment is conducted on a railway flatcar moving to the right, while *k* is that of a motorcar moving faster than the train in the same direction. In *k*′ the left measurement event *A* occurs before the right measurement event B; in *k* it occurs after B.

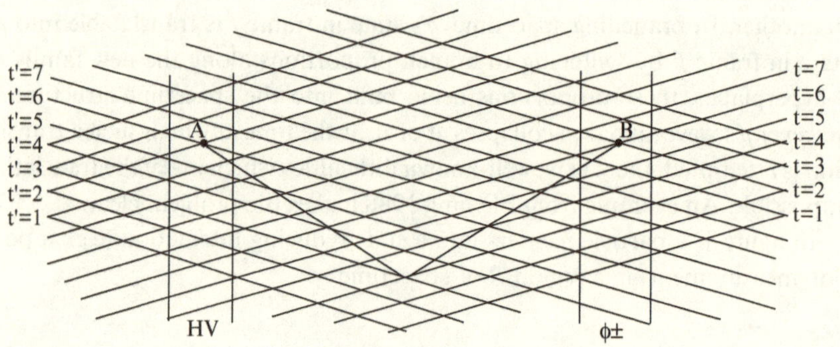

FIGURE 2

When emitted, the two-photon system is in the entangled spin state $|h\rangle|v\rangle + |v\rangle|h\rangle$. Each photon, considered separately, is in a state empirically indistinguishable from the mixed state $1/2(P_h + P_v)$, which is to say that each photon has a probability of 0.5 of passing + or − when its polarization is measured at any angle whatsoever. Suppose that the left polarizer is horizontal/vertical, and the right polarizer is oriented at an angle φ to the vertical. In frame *k*′, following the measurement of the left photon, the entangled state partially collapses and the right photon is thrown into a pure spin state. If the left outcome is *h*, the right photon enters the state $|v\rangle$, and its probability of being detected in the φ+ channel is $\cos^2 \varphi$. But if the left outcome is *v*, the right photon enters the state $|h\rangle$, and its probability of passing φ+ is $\sin^2 \varphi$.

Two questions arise:

(i) How can the right photon, at the moment it enters its analyzer, be both in a mixed state from the standpoint of frame *k*′, and in a pure state from the standpoint of *k*?

(ii) How, in frame *k*′, can the outcome of the left measurement event affect the state of the right photon, given that the two photons are receding from each other at the speed of light?

The first of these questions is discussed in the next section, and the second in Section 6.

4. Hyperplane Dependency and Covariant State Vector Reduction

Fleming's method of dealing with the apparently contradictory states of the right photon—mixed in frame k' and pure in frame k—is to introduce the concept of hyperplane dependency. We cannot speak simply of the "state" of a quantum system but always of its "state relative to a hyperplane". As Maudlin points out, we are familiar with relational states and properties: familiar with Socrates not being "tall" in any absolute or intrinsic sense, but only "tall relative to Theaetetus" (Maudlin 1994, 210). The question is, can polarization be a relational property of a photon, analogous to tallness? Or is polarization an intrinsic property, analogous to being snub-nosed?

Fleming (1996) defends the notion of hyperplane dependency without answering the last question explicitly, but Wayne (1997) supports the idea that the polarization states of photons in the EPR experiment are in an important sense relational rather than purely intrinsic. For Wayne, a single photon in a two-photon entangled system $|h\rangle|v\rangle + |v\rangle|h\rangle$ has "no intrinsic state: instead, the particle pair has an irreducibly relational state" (562). Just as, in STR, the spatial length of an object is a hyperplane dependent property, so the polarization state of each photon is determined by "hyperplane-dependent temporal relations between the photon and the measurement event at the distant detector" (565).

A useful notion in the present context is that of separability. Don Howard introduces the term to describe Einstein's belief in the independent existence of objects located in different regions of space. In Howard's definition physical systems in disjoint regions are "separable" if (a) each possesses its own distinct physical state, and (b) the joint state of the two systems is wholly determined by their separate states (Howard 1989, 226). Plainly the notion of hyperplane dependence violates separability. At the same time, in addition to separability, Einstein also believed in locality, namely that events in a region X of spacetime cannot be affected by events in spacelike separation from X.

The immediate task is to reconcile a description of EPR in one frame, including collapses, with its description in another frame. The great virtue of Fleming's pioneering work in this area is that it presents clearly, in a way that cannot be ignored, the problems of developing a theory of wavefunction collapse in a relativistic context. In the branched model, branching occurs along different crisscrossing hyperplanes in different frames. What happens to a photon in an entangled state when its twin is measured?

In a single spacetime manifold it is possible for the right photon to be in a mixed state in a given region, and it is possible for it to be in a pure state, but it is not possible for it to be both pure and mixed in the same region. In the branched model the situation is different. There are many future branches relative to the initial EPR conditions; on some of them the photon enters its analyzer in a mixed state, and in others in a pure state. Since the model branches along different hyperplanes in different frames, there will be sub-models in which the right measurement occurs respectively (i) before, (ii) after, and (iii) simultaneously with the left measurement. On branches in models of types (i) and (iii), the right photon enters its analyzer in a mixed state, and on branches of type (ii) it enters in a pure state. On its own branch each photon's polarization state is intrinsic to it. By putting different possible outcomes on different branches, depending on the frame in which the story of the experiment is unfolding, the model avoids making polarization states relational.

Summing up, figure 2 shows how, based on the branched model, consistent relativistic accounts of the EPR experiment can be given in frames k and k'. "Translation" from one to the other requires going to the submodels and reading off the respective histories. In neither case are we forced to describe a photon as being in two contradictory polarization states at the same time.

5. Lorentz-Invariant Probability Values

Across translations from one frame to another, probability values of the possible outcomes of a single quantum event remain invariant. The structure of the model which demonstrates this, and which also guarantees that every stochastic process has a unique outcome, is complex. What follows is a bare outline which may be omitted on first reading, although here as elsewhere "the devil is in the details".

Each "history", or complete path through the model, is a four-dimensional spacetime manifold. Every history contains a discrete set of "choice points", and the model branches along every "slice" (maximal set of space-like separated choice points). The spacelike hypersurfaces ("branching surfaces") defined by slices are partitioned into parallel families, each of which approximately fits the foliation of equal-time hyperplanes in a frame of reference. At a small interval Δt above each branching surface S are located the "outcomes" of its choice points, the probability of each outcome being determined by the proportion of branches which split along S and on which the outcome occurs.

Since each branching surface S belongs to one and only one parallel family, which in turn corresponds to a set of inertial frames which differ only minimally from one another, the probability of an outcome at a choice point is relative to the family to which S belongs. These probabilities (which reflect the lawlike character of QM) must be Lorentz-invariant. For the branched model to serve as a means of combining QM and STR it must therefore possess the following symmetry: for all parallel families, the proportions of branches on which a given type of outcome occurs above a given choice point are the same.

Through each choice point C in each history there pass many branching surfaces at different angles, on the branches above each of which are located C's possible outcomes. Branch attrition consists in the selection of one and only one branch B of these. Since B is unique, C has a unique actual outcome. Immediately after this outcome, B in turn splits along other branching surfaces containing B's choice points, which again have unique actual outcomes. The actual outcomes of all regions of spacetime are "accessible" to one another; they belong to the same history.

6. A Spacetime Mechanism for Nonlocal Effects

The most distinctive and unusual feature of the branched model is that it provides a mechanism which explains nonlocal connections: how it is possible for the outcomes of two or more stochastic, spacelike separated events to be correlated. In Section 3 it was asked how, in frame k', the outcome of the left measurement event in the EPR experiment could affect the state of the right photon, given that the photons are moving in opposite directions at the speed of light. Conversely, in frame k, how could the outcome of the right measurement affect the left photon? An answer to these questions provides the branched solution to the difficult puzzle of nonlocal connections.

Figure 3 pictures the dynamic behavior of the two-particle system in frame k'.

At A the entangled state partially collapses, the relative proportions of h- and v-branches being 0.5. If the outcome of the left measurement is v this means that a v-branch is stochastically selected (with probability 0.5), and on this branch the state of the right photon is $|h\rangle$. On the other hand if the left outcome is h then an h-branch is selected, and on this branch the state of the right photon is $|v\rangle$. In frame k', the selection of a single branch, brought about

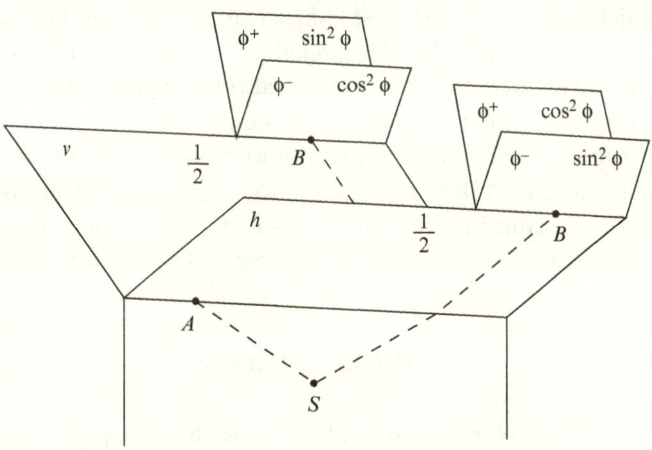

FIGURE 3

by the left measurement, produces an instantaneous effect on the right photon. The photon starts off in the mixed state $1/2(P_h + P_v)$, and is thrown into state $|h\rangle$ if the selected branch is a *v*-branch, and into $|v\rangle$ if the selected branch is an *h*-branch. Branch attrition along *k'*-hypersurfaces therefore accounts for the instantaneous influence (in frame *k'*) of the left measurement outcome on the polarization state of the right photon. This influence in turn generates the statistical correlations between the outcomes (*h,v*) on the left and (φ+,φ–) on the right. If the left outcome is *h*, then the probability of φ+ on the right is $\cos^2\varphi$, whereas if the left outcome is *v*, the probability of φ+ on the right is $\sin^2\varphi$.

In frame *k*, where the right measurement precedes the left, a mirror-image spacetime structure shows the state of the left photon being instantaneously influenced by the outcome of the right measurement. By the equivalence of inertial frames, neither of these frame-dependent histories is privileged over the other, and therefore there is no unidirectional causal relationship linking the two outcomes. (Causation is necessarily an asymmetric relation: if F is the cause of G, G cannot also be the cause of F.) Nevertheless, although not connected by causation proper, the two outcomes *are* connected. What connects them is a reciprocal, empirical, non-logical relationship of mutual dependence which is instantaneous in whatever frame one chooses to describe it. It's a "nonlocal connection" deriving from the way event-types are instantiated in branching spacetime, and from the fact that the branching extends along spacelike hypersurfaces.

In the description two paragraphs ago of the spacetime mechanism which gives rise to nonlocal effects, the concept of branch attrition played an important part. Thus the right photon is "thrown into state $|h\rangle$ if the selected branch is a *v*-branch". Certainly branch attrition is required if quantum experiments are to have unique outcomes: out of a range of possible outcomes it is branch attrition which selects the actual outcome. That being said, the existence of nonlocal connections can be read off, or read into, branching spacetime without branch attrition. In figure 3 the worldline of the right photon splits at the point at which it intersects the lowest *k'*-hypersurface. On some branches above that split the photon is polarized horizontally, whereas in others it is polarized vertically. But the lowest *k'*-hypersurface is also the hypersurface on which *A*, the left measurement event lies, and it is precisely on branches where *A* has outcome *v* that the right photon is polarized horizontally. Also, it is on branches where *A* has outcome *h* that the right photon is polarized vertically. All this information is contained in the branching structure, independently of branch attrition. Branch attrition is needed for collapse, but not for non-locality.

7. 3D/4D Equivalence

Throughout this chapter, the concept of branching has played the role of the Prince of Denmark in *Hamlet.* Different possible outcomes of stochastic processes are located on different branches; their probabilities are defined by proportionality relations among the branches on which they are located; quantum states are identified with such proportionalities in future branching structures, etc. However, the ontological burden imposed by branching spacetime is considerable, and there may be those who would prefer the coming together of QM and STR at lower cost. Some suggestions in this direction are contained in what follows.

3D/4D equivalence is a philosophical thesis which holds that to describe the world as a set of three-dimensional objects evolving in time, or alternatively as a set of four-dimensional objects extended in time, are equivalent. Every description of the first kind can be translated without loss into a description of the second kind, and vice versa (McCall 1994, 215–217). If the thesis of 3D/4D equivalence is correct, a dragonfly may be regarded either as three-dimensional object which moves jerkily, or equivalently as a four-dimensional worldline which bends erratically.

As was seen above, a stochastic process may be represented four-dimensionally in branching spacetime by putting each possible outcome on a separate branch, and allowing the actual outcome to be selected by branch attrition. The probability of that outcome is objective, defined by branch proportionality. In three dimensions the object undergoing the process behaves stochastically: the state it enters into at the end of the process is its actual state, and the possible but non-actual resultant states are not represented. For example, an electron passing through a Stern-Gerlach apparatus emerges in one channel or the other, and the outcome which might have resulted but didn't does not enter into the 3D description.

In four dimensions, what bestows a unique actual outcome upon a stochastic process is branching, combined with branch attrition. In order for the 3D and 4D descriptions to be perfectly equivalent, it is necessary that the probability of the actual outcome, in three dimensions, should be as objectively rooted in the real world as the probability value in four dimensions, based on branch proportionality. 3D/4D equivalence entails this, namely that in three dimensions the outcome of a probabilistic process is rigidly constrained by objective probabilities, the values of which are facts about the world in the same way that the atomic numbers of aluminum and copper are facts about the world.

Provided this last condition is met, as 3D/4D equivalence implies that it is, it makes no difference whether the world is described three-dimensionally or four-dimensionally. Seen in this light branching spacetime, and branch proportionality, may be treated as the ladder which is thrown away once we have climbed up the wall. In the context of quantum mechanics, and the reconciling of QM and STR, branching serves to give precise quantum probability values an ontological footing in the spacetime world. Those who wish to attach no more fundamental ontological significance to branching can simply assume that this has been done, and that the footing exists. Specifically, objective probability values (including joint probabilities of distant events) are established relative to spacelike hypersurfaces belonging to parallel families, each family representing a set of reference frames. Once this has been accomplished the ladder of spacetime branching can be thrown away, and the task of combining QM and STR proceeds within a single Minkowski manifold.

References

Albert, David Z. (1992). *Quantum Mechanics and Experience.* Cambridge, MA: Harvard University Press.

Bohm, David (1952). "A Suggested Interpretation of the Quantum Theory in Terms of 'Hidden' Variables. I and II". *Physical Review* 85: 166–179; 180–193.

Bohr, Niels (1949). "Discussion with Einstein on Epistemological Problems in Atomic Physics", in P. A. Schilpp (ed.). *Albert Einstein: Philosopher-Scientist.* LaSalle, IL: Open Court. 201–241.

Cushing, James T. (1994). *Quantum Mechanics.* Chicago: University of Chicago Press.

——, (ed.) (1996). *Bohmian Mechanics and Quantum Theory: An Appraisal.* Dordrecht: Reidel.

DeWitt, Bryce S. and Neill Graham (1973). *The Many-Worlds Interpretation of Quantum Mechanics.* Princeton: Princeton University Press.

Fleming, Gordon N. (1988). "Lorentz-Invariant State Reduction and Localization", in A. Fine and J. Leplin (eds.). *PSA 1988*, vol. 2: 112–126.

——. (1996), "Just How Radical Is Hyperplane Dependence?", in R. Clifton (ed.), *Perspectives on Quantum Reality*. Dordrecht: Reidel. 11–28.

Howard, Don (1989), "Holism, Separability, and the Metaphysical Implications of the Bell Arguments", in J. Cushing and E. McMullin (eds.). *Philosophical Consequences of Quantum Theory.* South Bend, IN: University of Notre Dame Press. 224–253.

Maudlin, Tim (1994). *Quantum Non-Locality and Relativity.* London: Blackwell.

——. (1996). "Spacetime in the Quantum World", in Cushing 1996, 285–307.

McCall, Storrs (1994). *A Model of the Universe.* New York: Oxford University Press.

——. (1995). "Time Flow, Non-Locality, and Measurement in Quantum Mechanics", in S. Savitt (ed.). *Time's Arrows Today.* New York: Cambridge University Press. 155–172.

Wayne, Andrew (1997). Review of [Maudlin 1994]. *Noûs* 31: 556–567.

{ 19 }

Downward Causation, Biological Information, and Development Fields

Storrs McCall and E.J. Lowe

In the "layered" conception of science, the bottom level is occupied by elementary particle physics, and above that lie solid state physics, chemistry, biology, psychology, etc. Each level possesses its own causal laws, and the concept of "downward causation" denotes possible causal influences from a higher to a lower level. We argue in this chapter that the downward causal influence of higher-level biological information upon molecular and other physical events in developing organisms is what constitutes "life", and that not all the needed information is stored in an organism's DNA.

1. The Hierarchical Structure of Science and the World

In 1958 a famous article by Paul Oppenheim and Hilary Putnam, "Unity of science as a working hypothesis", contained the vision of a single unified science covering everything that happens in the world:

> "It is not absurd to suppose that psychological laws may eventually be explained in terms of the behavior of individual neurons in the brain; that the behavior of individual cells—including neurons—may eventually be explained in terms of their biochemical constitution; and that the behavior of molecules—including the macro-molecules that make up living cells—may eventually be explained in terms of atomic physics." (Oppenheim and Putnam (1958), p. 7)

In the Oppenheim-Putnam world-view, science consists of only one layer, everything being reducible to particle physics. The alternative view, that there are many independent, mutually irreducible layers, is expressed amusingly in the title of J.A. Fodor's "Special sciences (or the disunity of science as a working hypothesis)" (1974). The multi-layer theory receives its classic exposition in a paper of the solid-state physicist Philip Anderson, entitled "More is different":

> It seems to me that one may arrange the sciences roughly linearly in a hierarchy, according to the idea: The elementary entities of science X obey the laws of science Y.

X	Y
solid state or many-bodied physics	elementary particle physics
chemistry	many-body physics
molecular biology	chemistry
cell biology	molecular biology
.	
.	
.	
psychology	physiology
social sciences	psychology

> But this hierarchy does not imply that science X is just applied Y. At each stage entirely new laws, concepts, and generalizations are necessary, requiring inspiration and creativity to just as great a degree as in the previous one. Psychology is not applied biology, nor is biology applied chemistry. (Anderson (1972), p. 222 of the (2008) reprint.)

When Anderson says that entirely new laws, concepts, and generalizations are found at each stage of the hierarchy, he is appealing to the notion of *emergence*, originally proposed by G.H. Lewes (1875) and made use of by Samuel Alexander (1920), Lloyd Morgan (1923), and C.D. Broad (1923). New laws, concepts, etc. emerge at each level. A good survey is Brian McLaughlin's "The rise and fall of British emergentism" (1992). In recent years the at times maligned notion of emergence has once more become popular, and two anthologies, Clayton and Davies' *The Re-Emergence of Emergence* (2006), and Bedau and Humphreys' *Emergence* (2008), contain numerous contemporary analyses. The fact that properties such as liquidity, solubility, life, and consciousness, make no sense in particle physics but emerge only at higher levels, does not preclude their being influenced by "upward causation". Causes at lower levels can have effects at higher levels. Nor is it far-fetched to suppose that some higher-level phenomena, such as thoughts and sensations, may "supervene" on lower-level neurological states. This means that two mental states cannot differ without a difference in brain states, although two brain states can differ without any difference in the corresponding mental states (see Kim [1993], pp. 54–57). But supervenience and "upward causation" are one thing; "downward causation" is another. The first use of the latter expression occurs in Campbell (1974); see the anthology *Downward Causation* (2000), ed. P.W. Anderson.

2. Downward Causation at the Biological Level

Upward causation and supervenience of higher-level states upon lower-level ones are consistent with the possibility of reducing all levels to physics. But downward causation is not. In the case of biology, if it can be shown that higher-level informational states influence the behaviour of lower-level cells and molecules, then biology will not be reducible to physics unless the higher-level informational states themselves are reducible. In what follows we shall argue that this is not the case, since biological development is guided by (i) information stored in nucleotide sequences in DNA and RNA, as well as (ii) other information, not so stored. Although (i) might be considered information "reducible" to physics, in the sense that nucleotide sequences are physical entities, information of type (ii) cannot be so reduced. Or so we shall try to show. It follows that biology is not overall reducible, and the category of "life" cannot be defined within physics alone.

3. Biological Information

The double helix structure of the DNA molecule is composed of the four nucleotide bases adenine (A), cytosine (C), guanine (G) and thymine (T). Sequences of these bases, e.g. CTGGATTT…CCA, code for the production of the proteins out of which living organisms are built. All genetic information is stored in DNA. In the words of George C. Williams, a pioneer in molecular genetics:

> DNA is the medium, not the message. A gene is not a DNA molecule; it is the transcribable information coded by the molecule…the gene is a packet of information, not an object. (Williams (1992), quoted in Sterelny and Griffiths (1999), p. 100)

In information theory, there are two forms which information can take: digital and analog. Digital information consists of a series of discrete inputs, for example "1"s and "0"s, or a nucleotide sequence in DNA, which results in a series of discrete outputs. In analog information, on the other hand, both input and output signals vary continuously, and there is no smallest "bit" of information. A photograph taken by a film camera yields analog information about a human face, or a landscape, whereas a digital camera produces thousands of tiny "pixels", the discrete differences between which our eyes are not able to detect. Genetic information is digital: either a given DNA molecule codes for the production of a given protein or it does not. There are no half-measures or approximations in this process. It is also irreversible. The "Central Dogma" of molecular biology, formulated by Francis Crick in 1957, asserts that:

> Once "information" has passed into protein, *it cannot get out again*. In more detail, the transfer of information from nucleic acid to nucleic

> acid, or from nucleic acid to protein may be possible, but transfer from protein to protein, or from protein to nucleic acid is impossible. (Quoted in Judson (1996), p. 332)[1]

John Maynard Smith remarks that the Central Dogma is one of the few statements in biology that is (i) non-trivial, (ii) not implied by the laws of physics, and yet is (iii) considered to be universally true (Maynard-Smith (2000b), p. 214). This fact argues strongly for emergence, and the "layered" conception of science.

The link between biological information and downward causation is provided by Paul Davies, who suggests that the difference between "information" in biology and what he calls "hardware" is parallel to the difference between "blueprint" for constructing protein and "molecular storage of the blueprint" in DNA and RNA. Davies says:

> In biology, genes are regarded as repositories of information—genetic databanks. In this case the information is semantic; it contains coded instructions for the implementation of an algorithm. So in molecular biology we have the informational level of description, full of language about constructing proteins according to a blueprint, and the hardware level in terms of molecules of specific atomic sequences and shapes. Biologists flip between these two modes of description without addressing the issue how information controls hardware (e.g. in "gene silencing" or transcription-inhibition)—a classic case of downward causation. (Davies (2006), p. 45)

As was mentioned above, information can be conveyed in digital or in analog form, and genetic information is conveyed in digital form. But is this the only form that biological information can take? Could the development of organisms also be guided by analog information? This is an important and interesting question, one that will occupy the remainder of the chapter, and is connected to the much-discussed topic of "morphogenesis" in biology, and the concept of a "morphogenetic field", or "development field". Morphogenesis, from the Greek word *morphe*, form or shape, is the process by which living beings, and the tissues, organs and structures that constitute them, develop and receive their form. The role that biological information plays in this process is key to understanding how organisms come into being, and how they differ from inanimate things.

4. Morphogenesis, and Morphogenetic Fields

Perhaps the most obvious fact about living individuals, and species of individuals, is that their growth and development is not random, but appears to proceed

[1] In fact, the view expressed in the "Central Dogma" is not strictly correct. Prions transfer information directly from protein to protein, and retroviruses modify DNA using RNA sequences.

according to a fixed pattern or plan. Sometimes there are abnormalities, but the fact of their being called "abnormal" reinforces the standard of a fixed plan that can on occasion be departed from. The question is, what guides the development of form: the morphogenesis of a new-born infant, a limb bud, or an embryonic eye into a fully formed adult entity?

Bruce Carlson, in his updated edition of Patten's *Foundations of Embryology* (1981), says the following:

> Many types of processes can contribute to the morphogenesis of a system. One of the most fundamental, but least understood, is *pattern formation*. There is evidence that very early in the development of many structures, prior to the onset of cell differentiation, an invisible pattern or blueprint is laid down and that further development uses this pattern as a guide. (Carlson (1981), p.17)

Carlson also remarks that:

> Morphogenesis remains one of the major mysteries of biology, and our present knowledge of this field is so slight that can be compared with the state of genetics before the rediscovery of Mendel's laws. (*Ibid*, p. 16)

Note the contrast, implicit in the above quotation, between "pattern formation", and "following a pattern". Between these two there is a sharp difference. As organisms develop, many writers speak of a pattern *forming*, or *emerging*. These words imply that before the biological pattern was formed, or emerged, *it did not exist*. But the phrase "following a pattern", or "following a blueprint" is very different in its logical import. It implies that the pattern or blueprint exists *before* the biological development. Plainly we cannot have it both ways. Either the pattern exists before the development begins, and informs and guides it, or the pattern emerges only once the development has taken place. Which is it to be?

The orthodox answer, in modern molecular biology, is that the *information* which leads to the pattern exists before development. That information is *genetic* information, stored in the organism's DNA. In an indirect sense, then, the pattern exists before development begins. For centuries, the same issue was debated between the supporters of *pre-formation* as opposed to *epigenesis* in biology. Aristotle was an epigenesist, and observed that freshly laid eggs contain no preformed chick, but that gradually a heart developed in the egg-white which began to beat, followed by the other parts that make up a chicken (*Hist. Animalium* 561a4–21). In the 17th century, Nicolaas Hartsoeker provided the image of a tiny perfectly formed homunculus curled up in the head of the sperm (reproduced in Wolpert (2002), p. 5). Although it is now universally accepted that the form of living creatures emerges over time, a vestige of pre-formationism exists in the encoding of a genetic blueprint for each developing embryo in that embryo's DNA. Valuable discussions of morphogenesis and pre-formation are found in Maienschein (2005), and in Huxley and de Beer (1934), pp. 1–8. What needs to be clarified is the precise manner in which biological information guides, shapes, and in conjunction with environmental factors causally determines the development of organisms.

In a short and compelling passage dealing with cell differentiation from the standpoint of molecular biology, James D. Watson and his co-authors present an essential key to understanding the processes by which genes determine embryonic development (Watson et al. (1987), p. 748). They begin by conceding that, in its details, the problem is complex beyond our powers of comprehension. In their words, "If each of the many hundreds of gene products that distinguish a liver cell from a muscle cell had to be studied in minute detail before we could understand the molecular causes of development, the situation would be virtually hopeless". However, for many or perhaps all developing organisms, genes can be arranged in a hierarchy. Some genes control the expression of other genes. Thus the myriads of genes that guide the development of a single cell "are not all equal, because they control the initial decisions to differentiate along one development pathway or another. The goal of molecular biologists studying development is therefore to discover which genes control the expression of other genes" (p. 749). (See also Wolpert (2002), pp. 13–17.) One of the problems faced by embryologists is that cells frequently become "committed" or "determined" to differentiate long before any actual morphological change is observed. In molecular biology, "determination" results from the expression of a controlling gene, while actual cell differentiation can take place substantially later.

Plainly, if genes in DNA have the ability to control growth and development as well as to control other genes, cells must be able to measure time. The ordering and timing of gene activation is critical. Watson et al. conjecture that all living beings have accurate clock mechanisms at the cellular level, while acknowledging that we have as yet little information about how such mechanisms function (p. 753). However, once the concept of a development field is accepted, the timing of events in biological development becomes more easily accounted for. What is required is that such fields be taken as extending in four dimensions, not only in three. In this respect a development field resembles an Einsteinian gravitational field in general relativity, which unlike a standard electrical or magnetic field is also four-dimensional. As the planets circle around the Sun, they follow geodesics (shortest paths) in curved four-dimensional spacetime. An interesting research project in field-morphogenesis would be to establish the shapes or volumes in four-dimensional development space that correspond to "freely developing" organisms. A freely developing organism (cf. "freely falling body") is an organism the development of which is not hampered by factors such as disease, or poisoning, or amputation of a limb.

A three-dimensional time-slice of a four-dimensional development field, that is to say the three-dimensional form at a given moment in time of an organism

X, specifies the state of X at that moment. Since the various stages in X's morphogenesis are mapped out along the time-axis of X's development field, there need be no mystery, once the field concept is accepted, of which events occur at which times. The "switching on" of a gene, or cell division in the blastula, or the growth of a limb-bud; all these events are recorded in finest detail in the 4D morphogenetic field. This is not to say that individual cells may not also possess some sophisticated 3D mechanism for time-keeping. But such a mechanism is not strictly speaking necessary, given morphogenetic fields, any more than the concept of gravitational attraction is necessary once the idea is accepted that the Earth follows a geodesic in four-dimensional spacetime around the Sun. Where appropriate, the conceptual simplification of moving to four dimensions is not to be underestimated.

5. Development Fields: Gurwitsch, Spemann, Weiss, Huxley, Needham

The first biologist to propose the idea of a morphogenetic or development field was the Russian Alexander Gurwitsch ("On the concept of an embryonic field" (1922)). At about the same time, Hans Spemann, Nobel Prize winner in 1935, was also exploring the field concept. Spemann describes his own work, together with that of Gurwitsch and Paul Weiss, in his text *Embryonic Development and Induction* (1938). Chapter 15 of Spemann's book is entitled "The embryonic field".

The biologist who contributed most to the conceptual analysis and understanding of development fields was Weiss, whose earliest work on the subject was contemporaneous with that of Gurwitsch and Spemann, and who articulates the theory of such fields in his *Principles of Development* (1939). On pp. 293–94 Weiss lists seven criteria that characterize development fields. These are concisely summarized by Joseph Needham in *Biochemistry and Morphogenesis* (1950):

> A field is (1) bound to a material substratum, (2) an entity, a pattern, not a mosaic, (3) heteroaxial and heteropolar, (4) able to include recognizably separate districts, (5) able to maintain its pattern when its mass is reduced, (6) able to perpetuate its pattern when completely divided into two or more pieces, (7) able to maintain its pattern when its mass is increased, and to fuse with similar pattern entering with the new material if the axial orientations are favorable (Needham p. 128, footnote)

For Weiss, development fields have an objective existence as part of the physical world:

> The fact that practically all developmental phenomena exhibit field-like characters in one or the other respect, is...a strong indication that the

> field concept is not only a useful circumlocution, but an expression of *physical reality*. (Weiss (1939), p. 292)

Being something physical, biological fields presumably are as open to study as electric, magnetic, and gravitational fields.

In Huxley and de Beer's *Elements of Experimental Embryology* (1934), chapters 8–10 are devoted to an extensive discussion of morphogenetic fields. The notion of a *field-gradient* is introduced as essential to understanding in detail the role of the field in development, for example development that concerns regeneration of limbs and other body parts in animals such as amphibia. The ability of a newt to regenerate an amputated limb, and more significantly the type of limb that results from such regeneration, appears to depend on a field-gradient. This gradient varies according to the distance away from the newt's body the original amputation was made. If a "proximal" (i.e. close to the body) stump of a fore-limb taken from a previous amputation is grafted onto a newt's hind-limb stump, the resulting regenerated member is a fore-limb. If the graft is of a "distal" fore-limb stump (taken from farther away), the result is a hind-limb (Huxley and de Beer, p. 307). A continuously varying gradient appears to be at work here, which depends on the distance away from the body the amputation of the stump is made. This gradient can result in the regeneration of either a fore-limb or a hind-limb, depending on the gradient's value. The continuously varying effect of the gradient on the newt's future bodily configuration is characteristic of the operation of a field, rather than the all-or-nothing expression of a gene. If a gene-expression explanation is demanded, the question arises, what switches on a fore-limb gene, as opposed to a hind-limb gene? The only possible answer seems to be, a field gradient of some sort.

What was not foreseen in the 1930s and 40s, the heyday of development field theory, was that the field concept of morphogenesis would disappear overnight once the digital coding of biological information in DNA was discovered. Opitz (1985) remarks:

> In one of the most astounding developments in Western scientific history, the gradient-field, or epimorphic field concept, as embodied in normal ontogeny and as studied by experimental embryologists, seems to have simply vanished from the intellectual patrimony of Western biologists. (Gilbert et al. (1996), p. 360)

Gilbert et al. ask: What destroyed the morphogenetic field? Their answer is, that nothing destroyed it. Rather, it was eclipsed by the rise of genetics with its alternative explanation of how development proceeds. "Since morphogenesis was subsumed in the larger category of gene expression, fields were not needed" (ibid, p. 360). The switching on and off of genes, resulting in synthesis at specific times and places of the proteins needed for growth and development, provided an apparently full and satisfactory explanation of how organisms grow, and how biological forms and patterns are actualized. The morphogenetic field was unnecessary.

Is the matter now closed? Has pattern formation been fully explained? We argue in the next section that this is not so, and that genetic explanations of growth and development must be supplemented by something else. The most promising candidate is a revitalized conception of the development field.

6. Two Kinds of Biological Information

A morphogenetic field is essentially an "information field", through which the genetic information stored in DNA is supplemented by other information, not so stored. The difference between the two lies in the fact that DNA information is digital, while the information stored in morphogenetic fields is analog. An argument could be made for the primacy of genetic information over morphogenetic information. If DNA-like molecules with encoded instructions for protein manufacture had not emerged in the deep past, there would have been no morphogenetic fields. This is true. But, at the same time, there is more to life than protein manufacture. There is manufacture of the right kind of protein, at the right place and at the right time. These essential modalities must be supplied, we claim, by the analog information stored in morphogenetic fields.

Gilbert, Opitz and Raff express what they call the "rediscovery" of morphogenetic fields in recent years as being based on "genetically defined interactions among cells", the limits of which can be established by homeotic genes ((1996), p. 367). Homeotic genes are genes that determine "which parts of the body form what body parts". This means that homeotic genes delimit the regions of the developing body that will eventually become a heart, or a lung, or a leg, or an eye. The key requirement, for a homeotic gene to do its work, is that the organ or body part, the development of which the gene initiates at a given moment, should be the *right* organ or part for that particular organism, located in the *right* area. What is the overall plan? Is there a blueprint? It isn't clear that genes alone can bootstrap their way to establishing an overall body plan, for the following reason.

Genes get activated in specific body regions, at specific times. The question arises, what triggers the activation of a gene? What turns it on, and what turns it off? The orthodox answer is, a regulatory gene. But a regulatory gene is just another gene. If gene A is activated by gene B, and gene B is activated by gene C, what activates gene C? Sooner or later, something different has got to enter into the equation. That "something different" cannot be yet another piece of digital information stored in DNA. A plausible hypothesis is that it is analog information, stored in a morphogenetic field.

If this hypothesis is correct, the development of every organism from its origin as a fertilized egg, or from the point it separates, amoeba-like, from its parent, is guided by a development field. The information the field supplies is analog, not digital, and it provides an overall plan for gene activation, cell division, and the production of proteins and other body materials at the right place and the right time. The functioning of DNA, and RNA, could be compared to a factory, which works to a plan. The plan or blueprint is supplied by the morphogenetic field, which influences bodily development by downward causation.

It may be asked, how is the morphogenetic field supposed to "communicate", as it were, with the developing organism, "telling" it how to grow? A possible answer is supplied by analogy with the four-dimensional gravitational field of general relativity. In general relativity, it has been said, "matter tells spacetime how to curve, and spacetime tells matter how to move" (J.A. Wheeler). The relationship between moving bodies and gravitational fields lies in the strict correlation between spacetime curvature and the paths of freely falling objects, such objects following shortest paths in spacetime. Similarly, in biology, a "freely developing" organism, not subject to lack of food, or poisoning, or other artificial restraints, follows the development path laid down for it by its four-dimensional morphogenetic field. The organism's "communication" with the field simply consists in the correlation between its growth-stages and the values of its field at different times. Organisms forced out of their normal development paths by starvation, or amputation, or disease, follow paths that approximate normal development as closely as circumstances permit.

Consider the following example of a developmental process, which we propose as being due to the action of a morphogenetic field. The process involves bilateral symmetry, e.g. the bilateral symmetry of a human being's arms. Our arms grew out of limb buds, which form in human embryos at a very early stage. Once an infant is born, its two arms are approximately the same length. After birth they continue to grow for many years. What guarantees that the left and right arms grow at the same rate? Why isn't my left arm six inches longer than my right arm? It's easy to say that a gene controls this, but not so easy to say exactly how the gene might do so. Call the gene X. What X has to do is balance the production of bone and tissue protein which lengthens the left arm with the production of similar protein which lengthens the right arm. But how does X accomplish this? It could do so if, for example, arm length came in three and only three sizes, "long", "medium", and "short". Then regulatory X-genes could always work in pairs, long-long, medium-medium and short-short. If an X-long gene were dispatched to control protein production in the right arm, an X-long gene would be dispatched to the left arm. Similarly for X-medium and X-short genes. It would be impossible (barring some kind of bizarre mutation) for an X-long gene to go the right arm and an X-medium gene to the left. The difficulty with this gene scenario is that arm length does not come in discrete sizes. Arm length can vary continuously from individual to individual, and even in the same individual there is an infinite number of possible arm lengths (within limits set by nutritional factors, etc.) that the individual can attain in the period from birth to adulthood. So, to repeat, what controls "sameness of arm length, right and left"? How does the left arm "know" how much the right arm is growing? Barring a regulatory gene, the only answer seems to be: a single, unified development field for the individual in question.

As a final example of biological field-action, consider an example found in Kirschner, Gerhart and Mitchison (2000). The organism they select is a large single-celled ciliated protozoan named *Stentor coerulus* which is shaped like a trumpet, can grow up to 1 mm in length, lives in water with its foot attached to a stone or other solid object, and feeds by using cilia to sweep smaller organisms into its funnel-shaped gullet. Tartar (1961) is a full-length study of *Stentor*. Tartar was fascinated by the protozoan's ability to recover from surgical operations, and Kirschner et al. select *Stentor* as a prime example of what they describe as a "self-organizing mechanism", a mechanism which differs fundamentally from anything describable as a "machine", operating under the influence of contact forces and efficient causes. We argue that what makes *Stentor* and other living organisms "self-organizing" is not that they are mechanisms of a special kind, but that they grow and re-generate by following patterns laid down by a morphogenetic field.

Tartar's experiments performed on *Stentor* are remarkable. Using tiny scalpels, or scissors, he was able to cut the organism into small fragments, some only a thousandth of the cell's original volume. Provided a fragment contained part of the spiral fold of the cell's surface which forms the gullet (Spemann's "organizer"), within a day the fragment was able to re-organize itself into a smaller replica of the original animal. These mini-*Stentors* gradually enlarged as they fed (Kirschner et al., p. 83). For "prolonged survival" (Kirschner does not say how long) the original cell's nucleus was necessary. The question is, how does *Stentor* do it? What "principle of re-organization" brings about the transformation of the tiny fragment of *Stentor* into a tiny replica of the original animal?

If *Stentor* were a multi-celled creature undergoing cell division, the answer might be that all the information needed for re-generation of the animal's original form was contained in the DNA of the surviving cell or cells. In that case, the re-organization of the mini-*Stentor* from the surviving fragment would be a "genomic" phenomenon based on the genes encoded in the organism's DNA molecules. But *Stentor* is a single-cell protozoan, and does not grow by cell division. As Kirschner et al. remark, "We know very little about how *Stentor*, or any other ciliate, self-organizes a specific body plan". "How the circumferential axis in *Stentor* is generated, and how the LSC works [the "fold" in *Stentor*'s cell wall], are completely mysterious" ((2000), p. 83). They go so far as to say that we need to "move beyond the genomic analysis of protein and RNA components of the cell (which will soon become a thing of the past)" (p. 87). So, if genomic analysis cannot provide the source of the information needed for *Stentor*'s re-organization plan following surgery, what *can* provide it?

The suggestion made here is that the information guiding *Stentor*'s re-organization is not genetic, and requires no activation of *Stentor*'s genes. Immediately following surgery, *Stentor* need synthesize no new proteins using DNA and transcription RNA. The information *Stentor* needs to re-organize, we hypothesize, is contained not in DNA, but in the *Stentor* morphogenetic field, which in conformity with Weiss's criterion (6), extends intact around each viable fragment created by surgery. It can, and does, influence how bits of matter move and re-organize themselves into a mini-*Stentor*. The causal efficacy of the morphogenetic information field is an instance of downward causation, one of the defining characteristics of life.

References

Alexander, S. (1920), *Space, Time, and Deity*, New York: Humanities Press.

Andersen, P.B. et al. (eds) (2000), *Downward Causation*, Oakville, CT: Aarhus University Press.

Anderson, P.W. (1972), "More is different", *Science 177*, 393–96. Reprinted in Bedau and Humphreys (2008), 221–29.

Bedau, M.A. and Humphreys, P. (2008), *Emergence: Contemporary Readings in Philosophy and Science*, Cambridge, MA: MIT Press.

Broad, C.D. (1923), *Scientific Thought*, New York: Harcourt, Brace, & Co.

Campbell, D.T. (1974), "'Downward Causation' in hierarchically organized biological systems", in *Studies in the Philosophy of Biology*, ed. Ayala and Dobzhansky. Berkeley: University of California Press.

Carlson, B.M (1981), *Patten's Foundations of Embryology*.

Clayton, P. and Davies, P.C.W. (2006), *The Re-Emergence of Emergence*, Oxford: Oxford University Press.

Davies, P.C.W. (2006), "The physics of downward causation", in Clayton and Davies, 35–51.

Fodor, J.A. (1974), "Special sciences (or the disunity of science as a working hypothesis)", *Synthese 27*, 97–115.

Gilbert, S.F., Opitz, J.M. and Raff, R.A. (1996), "Resynthesizing evolutionary and developmental biology", *Developmental Biology 173*, 357–72.

Gurwitsch, A. (1922), "On the concept of an embryonic field".

Huxley, J.S., and de Beer, G.R. (1934), *The Elements of Experimental Embryology*, New York: Hafner Pub. Co.

Judson, H.F. (1996), *The Eighth Day of Creation: Makers of the Revolution in Molecular Biology* (Expanded edition), New York: Simon and Schuster.

Kim, J. (1993), *Supervenience and Mind*, New York: Cambridge University Press.

Kirschner, M., Gerhart, J. and Mitchison, T. (2000), "Molecular 'vitalism'", *Cell 100*, pp. 79–88.

Lewes, G.H. (1875), *Problems of Life and Mind*, Boston: Houghton, Osgood.

Lloyd Morgan, C. (1923), *Emergent Evolution*, London: Williams and Norgate.

Maienschein, J. (2005), "Epigenesis and Preformationism", *Stanford Encyclopedia of Philosophy*.

Maynard Smith, J. (2000b), "Reply to commentaries", *Philosophy of Science* 67, 214–18.

McLaughlin, B.P. (1992), "The rise and fall of British emergentism", in *Emergence or Reduction?* (ed. Beckerman, Flohr and Kim), reprinted in Bedau and Humphreys (2008), 19–59.

Needham, J. (1950), *Biochemistry and Morphogenesis*, Cambridge: Cambridge University Press.

Opitz, J.M. (1985), "The development field concept", *American Journal of Medical Genetics 2*, 1–11.

Oppenheim, P. and Putnam, H. (1958), "Unity of science as a working hypothesis", *Minnesota Studies in the Philosophy of Science II*, 3–36.

Spemann, H. (1967), *Embryonic Development and Induction*, New York: Hafner.

Sterelny, K. and Griffiths, P.E. (1999), *Sex and Death: An Introduction to Philosophy of Biology*, Chicago: University of Chicago Press.

Tartar, V. (1961), *The biology of Stentor*, Oxford: Pergamon.

Watson, J.D., Hopkins, N.H., Roberts, J.W., Steitz, J.A., and Weiner, A.M. (1987), *Molecular Biology of the Gene*, 4th edition, Menlo CA: Benjamin/Cummings.

Williams, G.C. (1992), *Natural Selection: Domains, Levels and Challenges*, New York: Oxford University Press.

Weiss, P. (1939), *Principles of Development*, New York: H. Holt and Co.

Wolpert, L. (2002), *Principles of Development*, Oxford and New York: Oxford University Press.

{ 20 }

The Supervenience of Truth: Freewill and Omniscience

Since the time of St Augustine in the fifth century, philosophers and theologians have debated the question of whether human freedom is consistent with God's omniscience. So exhaustive and painstaking has been the examination of this issue, it might seem out of the question that a new, previously unnoticed consideration could influence the debate. Precisely this, nevertheless, is the contention of the chapter. That truth is a supervenient quality has to the best of the author's knowledge not yet been proposed,[1] but, once acknowledged, it radically alters the terms of the discussion and makes it relatively simple to reconcile human free will with divine omniscience.

1. The Supervenience of Truth

The domain of interest is empirical truth, the truth and falsehood of propositions about the world of space and time, not propositions in the realm of mathematics, logic, or ethics. All empirical propositions, if true, are made true by events or states of affairs with dates and spatial locations, which may be in the past, the present, or the future. Consider the set T of all true dated, tenseless empirical propositions, e.g. atomic propositions of the form "Universal *e* is instantiated at spacetime location (x,y,z,t)", or "Event A occurs later than event B in coordinate frame *f*", together with all entailed truth functions and quantifications of these. Such propositions, if true at all, are true *simpliciter*, i.e. not at a time but timelessly. The thesis of truth supervenience states that T supervenes upon the spacetime universe, the "Minkowski world" consisting of

[1] This is not quite right. John Bigelow, in Bigelow (1988: 132), advances the thesis that "truth supervenes on being". He does not however discuss the problem of the *reciprocal* co-variance of truth and being. See section 1 below.

the set E of all spacetime events. In the language of supervenience, E comprises the *base elements* on which the set T of *supervening elements* supervenes.[2]

If indeed truth supervenes upon events, its doing so differs from the way mental states are said to supervene upon brain states, or moral qualities upon natural qualities. Normally, a defining characteristic of supervenience is that a difference at the supervening level necessarily implies a difference at the base level but not conversely: different sets of base elements may correspond to one and the same supervening element. In the case of truth this is not so. Every difference in the elements of T necessarily implies a difference in E, and conversely every difference in E necessarily implies a difference in T. If Columbus had sailed south instead of west the members of T would have been different from what they are, and if "Antony wooed Cleopatra" had been false instead of true the history of the world would have unfolded differently. Unlike moral qualities, true propositions co-vary in a reciprocal way with the events and states of affairs on which they supervene. Any change in one necessitates a change in the other.

Despite reciprocal covariance, the supervenience of truth upon events is an asymmetric relation. The asymmetry lies in *dependence*: the truth of true empirical propositions depends on what occurs in space and time, but what occurs in space and time does not depend on what propositions are true. It is not easy to say exactly what this supervenience-dependence consists in.[3] It seems plain that it is neither straightforwardly causal, nor logical, nor nomological dependence. I shall confine myself to flagging instances of asymmetric supervenience-dependence with words like "because", "since", or "the reason why". Thus "snow is white" is true *because* snow is white, but the reason why snow is white is not because the proposition "snow is white" is true. The reason why Imogen is in pain is that her C-fibres are activated, it's not that her C-fibres are activated because she's in pain.

To sum up, the truth of true empirical propositions supervenes on events, but events do not supervene on true propositions. This is so despite the fact that T co-varies reciprocally with E. True propositions "float above" the physical world in a sense in which the world does not float above true propositions. We turn now to the problem of the incompatibility of God's omniscience with human freedom.

2. Free Will and Omniscience

In this chapter, the type of omniscience considered is *timeless* omniscience. God is considered to exist *atemporally*, outside the framework of time, rather

---

[2] See Jaegwon Kim 1993 and 2005. A good survey of supervenience theories is Brian McLaughlin 1995.

[3] An analogue might be the sense in which Xanthippe's becoming a widow depends on Socrates's drinking the hemlock, and in which Socrates's drinking the hemlock does *not* depend on Xanthippe's becoming a widow.

than *sempiternally*, i.e. at every moment. Augustine's and Aquinas's conception of God is that he stands outside time rather than within it, and cognizes the world *sub specie aeternitatis*, not in a successive manner from one instant to another. The atemporal omniscience of God accords with the tenseless, temporally definite propositions that are the object of that knowledge. Thus if God is omniscient God knows, not the truth of the future-tense proposition "The sun will rise in 1000 years", but (assuming it is true) "The sun rises (tenselessly) on Oct 28, 3010". To say that God is omniscient is to say that God knows of every true proposition that it is true, and of every false proposition that it is false. Included in what an omniscient God knows is the unchanging set T of true empirical propositions.

A common version of the traditional argument[4] for the alleged incompatibility of freedom and omniscience takes the following form. Suppose at noon today I have a choice, or seem to have a choice, whether to stoop and tie my shoelace. Suppose also that God is omniscient. Now if I do tie my shoelace at noon, the proposition

(1) S.McC. ties his shoelace at noon, Oct 28, 2010

is true, and God knows it to be true. Was it possible for me to refrain from tying it? Well, if I did refrain, (1) would be false, and the negation of (1) would be true. In refraining I would appear to confute an item of God's knowledge, and bring it about that God fails to know something true. But that is absurd. An omniscient being can entertain no false propositions, and be unaware of no true ones. Therefore, on the assumption that I do in fact tie my shoelace it follows that I cannot refrain from tying it. Similarly if I do in fact refrain from tying it, God's omniscience implies that I cannot tie it. Either way, omniscience appears to deprive me of the ability to do anything other than what I actually do. It would be false to say before I act at noon today that I have the power to tie my shoelace *and* that I have the power not to tie it. If being free consists in possessing "two-sided" abilities of this kind, the power to do X and the power not to do X, then divine omniscience would seem to be incompatible with such freedom.

3. The "Supervenience" Resolution

Putting the argument of the previous section in more abstract form, imagine a situation in which God knows that x φ's, and ask if it is possible for x not to φ. In formal terms, where "G*p*" symbolizes "God knows that *p*", can the propositions $G\varphi(x)$ and $\diamond \sim \varphi(x)$ both be true? In modal semantics the answer is yes, as

[4] John Martin Fischer's anthology 1989 contains a valuable introduction and many references. See also Calvin Normore 1985 and Fischer 1992.

is shown by a simple two-world model, where possible world w_2 is "accessible from", or "possible relative to", world w_1. Worlds w_1 and w_2 might for example be regarded as worlds sharing a common past up to an instant t, and diverging thereafter.

$$\begin{array}{cc} w_1 & w_2 \\ G\varphi(x) & \\ \Diamond\sim\varphi(x) \ldots\ldots\ldots & \sim\varphi(x) \\ \varphi(x) & G\sim\varphi(x) \end{array}$$

What this model indicates is that God's knowing that x φ's in world w_1 does not constrain x, or limit its possibilities. Although x φ's in w_1, the supposition that it is possible for x not to φ is perfectly consistent, as is shown by the existence of world w_2, where x fails to φ and, by supervenience, God knows that x fails to φ. The idea that God's knowing that $\varphi(x)$ should somehow limit what x can or can't do runs counter to the supervenience of truth, for what God knows depends on what is true, and what is true depends in turn on what obtains, including what capacities obtain, in the domain of events. To suppose the opposite, and maintain that the way the world is depends on what is true, or on what God knows, violates supervenience by reversing the dependence-relation. To repeat, truth depends on events, and events do not depend on truth. I can freely tie my shoelace at noon, and as a result of my free act, God knows from all eternity that I tie it. Truth-supervenience implies the compatibility of free will with omniscience.

The principle behind the supervenience of truth seems to have been grasped by at least one Scholastic philosopher, Luis de Molina. In his *Concordia Liberi Arbitrii* of 1588 Molina writes:

> It was not that since God foreknew what would happen from those things which depend on the created will that it would happen; but, on the contrary, it was because such things would happen through the freedom of the will, that He foreknew it; and that He would foreknow the opposite, if the opposite was to happen.[5]

In this passage Molina states the essence of the asymmetric dependence-relation which lies at the basis of supervenience, although in this case Molina is talking about omniscience rather than truth. For Molina, God knows what happens in the world *because it happens*, rather than the other

[5] Translated from Molina's *Concordia* by Mourant (1954: 426). The Latin text reads "... *non quia Deus praescivit hoc esse futurum, ex iis quae ab arbitrio creato pendent, ideo futurum esse; sed e contrario, quia ita pro arbitrii libertate erat futurum, Deum id praescivisse; praesciturus oppositum, si oppositum... foret futurum.*" (*Concordia Liberi Arbitrii*, Paris: Lethielleux, 1876, p. 547)

way round. In parallel fashion, in the case of truth, a proposition stating "An event Q happens" is true because Q happens, rather than the event happening because the proposition is true. Molina's principle, which could be described loosely as "supervenience of God's knowledge", is derivable from the supervenience of truth if omniscience is defined as it is above: the set of truths known by God is identical with the set of all true propositions. The dependence relation necessary for supervenience is then derivable from the corresponding relation for truth-supervenience as follows:

(2) The reason why God knows that X does A is that "X does A" is a true proposition (definition of omniscience)

(3) The reason why "X does A" is a true proposition is that X does A (truth-supervenience).

From (2) and (3) it follows that:

(4) The reason why God knows that X does A is that X does A.[6]

To this we might add, with Molina, that if X does A through the exercise of freedom of the will, then God knows that (X does A through the exercise of freedom of the will). The compatibility of freedom with omniscience is established.

References

Bigelow, J. 1988. *The Reality of Numbers.* New York: Oxford University Press.

Fischer, J.M. 1989. *God, Foreknowledge, and Freedom.* Stanford: Stanford University Press.

Fischer, J.M. 1992. Recent work on God and freedom. *American Philosophical Quarterly 29*: 91–109.

Kim, J. 1993. *Supervenience and Mind.* New York: Cambridge University Press.

Kim, J. 2005. *Physicalism, or Something near enough.* Princeton: Princeton University Press.

McLaughlin, B. 1995. Varieties of supervenience. In *Supervenience: New Essays*, ed. Savellos and Yalcin. New York: Cambridge University Press.

Mourant, J.A. 1954. *Readings in the Philosophy of Religion.* New York: Crowell.

Normore, C. 1985. Divine omniscience, omnipotence and future contingents: an overview. In *Divine Omniscience and Omnipotence in Medieval Philosophy*, ed. T. Rudavsky. Dordrecht: D. Reidel Pub. Co.

---

[6] To be fully rigorous, the inference requires a principle of transitivity for reasons:

(5) If the reason for *p* is *q*, and if the reason for *q* is *r*, then the reason for *p* is *r*.

For example (Conan Doyle) if the reason why X was able to enter the stable unnoticed was that the dog did not bark, and if the reason why the dog did not bark was that X was the dog's owner, then the reason why X was able to enter the stable unnoticed was that he was the dog's owner.

{ 21 }

An Insoluble Problem

In a paper published in 1986, Michael Dummett envisages the following situation.[1] A fifth-rate but conceited artist living the 20th century is visited by a time traveler from the 21st century, an art critic who is impressed by those of the artist's works that have survived and have given him the reputation of being one of the greatest painters of his time. When the critic sees the works, he is disappointed by their quality, and realizes that the artist has yet to create the magnificent paintings that subsequently made him famous. The critic produces a portfolio of reproductions of the later works, but shortly after has to depart at short notice, being permitted for some reason to remain only for a limited time in the past. The artist manages to conceal the portfolio, and spends the rest of his life meticulously copying on canvas the reproductions left behind. These copies form the basis of his subsequent artistic renown.

Is there a problem here? Yes. The whole subject of time travel is rife with problems. As Kurt Gödel showed in 1949, there exist solutions to Einstein's equations of general relativity that permit the existence of closed timelike loops, along which time travelers could, in theory, travel.[2] It goes without saying that such travel cannot, without contradiction, permit the occurrence of events that would negate the very possibility of the voyage itself. For instance, a traveler who visits the past could not kill his or her own grandmother. On this see Lewis (1976). However, barring such limitations, time travel is physically possible. The critic does not *change* the past in any way, though he may be said to *influence* it (Horwich, p. 116). Nothing prevents the art critic from visiting the artist, nor the artist from copying the works that make him famous. The puzzle lies not in this, but in finding where artistic creativity enters the equation.

---

[1] See Dummett (1986), p. 155. I am indebted to Friedel Weinert for drawing my attention to this puzzle.

[2] A picture of such a loop may be found in Horwich (1987), p. 113.

No one doubts the aesthetic value of the artist's paintings, nor the sense in which the critic's reproductions reflect this value. But the aesthetic value of a work of art, as distinct from a natural phenomenon like a sunset or a butterfly's wing, lies in the artistic creativity that produces it. In the story of the artist and the art critic, where is the artistic creativity? A skilled technician, wielding a paintbrush, can copy a Michelangelo or a Picasso. But what the copying process lacks is Michelangelo's creativity.

This is the true puzzle that underlies Dummett's imaginative example. Given the right circumstances, time travel is possible. There are no conceptual obstacles to understanding (i) the critic's visit to the artist, nor (ii) how the artist paints the copies. What is incomprehensible is (iii) who or what creates the works that future generations value? Where is the artistic creativity to be found? Unlike the traditional "paradoxes of time travel", this problem has no solution.

References

Dummett, M. 1986. Causal loops. In *The Nature of Time*, ed. Flood and Lockwood: 135–69. Oxford and New York: Blackwell.

Gödel, K. 1949. An example of a new type of cosmological solution of Einstein's field equations of gravitation. *Reviews of Modern Physics* 21: 447–50.

Horwich, P. 1987. *Asymmetries in Time.* Cambridge, MA: MIT Press.

Lewis, D. 1976. The paradoxes of time travel. Reprinted in Lewis' *Philosophical Papers*, vol. II: 67–80. New York: Oxford University Press, 1986.

{ INDEX }